Testing Safety-Related Software

Springer
*London
Berlin
Heidelberg
New York
Barcelona
Hong Kong
Milan
Paris
Santa Clara
Singapore
Tokyo*

Stewart Gardiner (Ed.)

Testing Safety-Related Software

A Practical Handbook

With 39 Figures

Springer

Stewart N. Gardiner, BSc, PhD, CEng
Ernst and Young, Management Consultants, George House,
50 George Square, Glasgow, G2 1RR, UK

No representation or warranty, express or implied, is made or given by or on behalf of the editor and the contributors or any of their respective directors or affiliates or any other person as to the accuracy, completeness or fairness of the information, opinions or feasibility contained in this book and this book should not be relied upon by any third parties who should conduct their own investigations of this book and the matters set out herein and furthermore no responsibility or liability in negligence or otherwise is accepted for any such information, opinions or feasibility and the editor and contributors shall not be liable for any indirect or consequential loss caused by or arising from any such information, opinions or feasibility being loss of profit and/or loss of production. The contributors are: BAeSEMA Limited, G P Elliot Electronic Systems Limited, Lloyd's Register, Lucas Aerospace Limited, Rolls Royce Industrial Controls Limited, Nuclear Electric Limited, Rolls Royce plc, Scottish Nuclear Limited and The University of Warwick.

ISBN 1-85233-034-1 Springer-Verlag London Berlin Heidelberg

British Library Cataloguing in Publication Data
Testing safety-related software : a practical handbook
  1. Computer software - Testing 2. Computer software-
  Reliability 3. Industrial safety - Computer programs
  I. Gardiner, Stewart
  005.1'4
ISBN 1852330341

Library of Congress Cataloging-in-Publication Data
Testing safety-related software : a practical handbook / Stewart
  Gardiner, ed.
    p.   cm.
  Includes bibliographical references and index.
  ISBN 1-85233-034-1 (pbk. : alk. paper)
    1. Computer software — Testing.   2. Industrial safety — Software -
- Testing..   I. Gardiner, Stewart, 1945-   .
  QA76.76.T48T476   1998   98-34276
  620.8'6'028514—dc21   CIP

Apart from any fair dealing for the purposes of research or private study, or criticism or review, as permitted under the Copyright, Designs and Patents Act 1988, this publication may only be reproduced, stored or transmitted, in any form or by any means, with the prior permission in writing of the publishers, or in the case of reprographic reproduction in accordance with the terms of licences issued by the Copyright Licensing Agency. Enquiries concerning reproduction outside those terms should be sent to the publishers.

© Springer-Verlag London Limited 1999
Printed in Great Britain

The use of registered names, trademarks etc. in this publication does not imply, even in the absence of a specific statement, that such names are exempt from the relevant laws and regulations and therefore free for general use.

The publisher makes no representation, express or implied, with regard to the accuracy of the information contained in this book and cannot accept any legal responsibility or liability for any errors or omissions that may be made.

Typesetting: Camera ready by editor
Printed and bound at the Athenæum Press Ltd., Gateshead, Tyne and Wear
34/3830-543210 Printed on acid-free paper

Acknowledgements

This book is based upon technical reports produced by a number of authors and is based upon the results of a collaborative research project (CONTESSE) to which many persons contributed. The project was partly funded by the UK Department of Trade and Industry (DTI) and the Engineering and Physical Sciences Research Council (EPSRC) and was carried out between 1992 and 1995.

The editor gratefully acknowledges the following contributions to the book:

Preparation of the final book:

        K. Czachur (Lloyds' Register)
        K. Khondar (University of Warwick)
        R. Lowe, M. Mills and A. MacKenzie
                (BAeSEMA)

Authors of the technical reports that form the basis of the book:

BAeSEMA	Dr. S. Gardiner
	R. Lowe
	A. MacKenzie
	H. Morton
	S. Thomson
G P-Elliot Electronic Systems	M. Ashman
	K. Homewood
	D. Marshall
	A. Smith
	J. Tennis
Lloyd's Register	S. Allen
	K. Czachur
	K. Lee
	Prof. C. MacFarlane
	(University of Strathclyde)
	B. McMillan (University of Strathclyde)
	R. Melville (Brown Brothers)

Lucas Aerospace	A. Ashdown
	D. Hodgson
NEI Control Systems	Dr. J. Parkinson
	A. Rounding
Nuclear Electric	Dr. I. Andrews
	G. Hughes
	D. Pavey
	Prof. P. Hall (The Open University)
	Dr. J. May (The Open University)
	Dr. H. Zhu (The Open University)
	Dr. A. D. Lunn (The Open University)
Rolls-Royce	M. Beeby
	T. Cockram
	S. Dootson
	N. Hayes
	Dr. J. Kelly
	P. Summers
	Dr. E. Williams
	Prof. A. Burns (University of York)
	Dr. D. Jackson (Formal Systems(Europe))
	Dr. M. Richardson (University of York)
Scottish Nuclear	I. O'Neill
	P. Pymm
The University of Warwick	Dr. F. Craine
	Prof. J. Cullyer
	K. Khondkar
	Dr. N. Storey

The helpful advice of Tony Levene, the Project Monitoring officer for the DTI, is acknowledged.

Crown Copyright is reproduced with the permission of the Controller of Her Majesty's Stationery Office.

Extracts from IEC standards are reproduced with permission under licence no. BSI\PD\1998 1028. Complete editions of the standards can be obtained by post through national standards bodies.

Camera-ready copy was created by Syntagma, Falmouth.

>Stewart Gardiner – CONTESSE project manager
>(Now with Ernst and Young, Management Consultancy Services.)

July 1998

Contents

1	**Introduction**	**1**
1.1	Context	1
1.2	Audience	2
1.3	Structure	3
1.4	Applicable Systems	4
1.5	Integrity Levels	5
1.6	Typical Architectures	5
1.7	The Safety Lifecycle and the Safety Case	17
1.8	Testing Issues across the Development Lifecycle	18
1.9	Tool Support	22
1.10	Current Industrial Practice	23
1.11	The Significance Placed upon Testing by Standards and Guidelines	29
1.12	Guidance	31
2	**Testing and the Safety Case**	**33**
2.1	Introduction	33
2.2	Safety and Risk Assessment	34
2.3	Hazard Analysis	35
2.4	The System Safety Case	41
2.5	Lifecycle Issues	45
2.6	Guidance	54
3	**Designing for Testability**	**59**
3.1	Introduction	59
3.2	Architectural Considerations	61
3.3	PES Interface Considerations	62
3.4	Implementation Options and Testing Attributes	64
3.5	Software Features	76

3.6	Guidance	82

4 Testing of Timing Aspects — 83

4.1	Introduction	83
4.2	Correctness of Timing Requirements	84
4.3	Scheduling Issues	86
4.4	Scheduling Strategies	89
4.5	Calculating Worst Case Execution Times	94
4.6	Guidance	100

5 The Test Environment — 101

5.1	Introduction	101
5.2	Test Activities Related to the Development of a Safety Case	102
5.3	A Generic Test Toolset	104
5.4	Safety and Quality Requirements for Test Tools	107
5.5	Statemate	110
5.6	Requirements and Traceability Management (RTM)	113
5.7	AdaTEST	116
5.8	Integrated Tool Support	121
5.9	Tool Selection Criteria	121
5.10	Guidance	123

6 The Use of Simulators — 125

6.1	Introduction	125
6.2	Types of Environment Simulators	126
6.3	Use of Software Environment Simulation in Testing Safety-Related Systems	128
6.4	Environment Simulation Accuracy and its Assessment Based on the Set Theory Model	132
6.5	Justification of Safety from Environment Simulation	139
6.6	Guidance	141

7 Test Adequacy — 143

7.1	Introduction	143
7.2	The Notion of Test Adequacy	143
7.3	The Role of Test Data Adequacy Criteria	144
7.4	Approaches to Measurement of Software Test Adequacy	147
7.5	The Use of Test Data Adequacy	152
7.6	Guidance	154

8 Statistical Software Testing 155
- 8.1 Introduction 155
- 8.2 Statistical Software Testing and Related Work 156
- 8.3 Test Adequacy and Statistical Software Testing 157
- 8.4 Environment Simulations in Dynamic Software Testing 159
- 8.5 Performing Statistical Software Testing 160
- 8.6 The Notion of Confidence in Statistical Software Testing 166
- 8.7 Criticisms of Statistical Software Testing 167
- 8.8 The Future of Statistical Software Testing 168
- 8.9 Guidance 170

9 Empirical Quantifiable Measures of Testing 171
- 9.1 Introduction 171
- 9.2 Test Cost Assessment 171
- 9.3 Test Regime Assessment 177
- 9.4 Discussion of Test Regime Assessment Model 188
- 9.5 Evidence to Support the Test Regime Assessment Model 191
- 9.6 Guidance 194

References 195

Appendix A Summary of Advice from the Standards 201

Bibliography 213

Index 219

Chapter 1
Introduction

As software is very complex, we can only test a limited range of the possible states of the software in a reasonable time frame.

In 1972, Dijkstra [1] claimed that 'program testing can be used to show the presence of bugs, but never their absence' to persuade us that a testing approach alone is not acceptable. This frequently quoted statement represented our knowledge about software testing at that time, and after over 25 years intensive practice, experiment and research, although software testing has been developed into a validation and verification technique indispensable to software engineering discipline, Dijkstra's statement is still valid.

To gain confidence in the safety of software based systems we must therefore assess both the product and the process of its development. Testing is one of the main ways of assessing the product, but it must be seen, together with process assessment, in the context of an overall safety case. This book provides guidance on how to make best use of the limited resources available for testing and to maximise the contribution that testing of the product makes to the safety case.

1.1 Context

The safety assurance of software based systems is a complex task as most failures stem from design errors committed by humans. To provide safety assurance, evidence needs to be gathered on the integrity of the system and put forward as an argued case (the *safety case*) that the system is adequately safe. It is effectively impossible to produce software based systems that are completely safe, as there will always be residual errors in the software which have the potential to cause failures that cause hazards. It is also important to build systems that are affordable. A judgement needs to be made on the cost incurred in ensuring that the system is safe.

Testing is one of the main ways of assessing the integrity (safety) of a software based system. Exhaustive testing of the software is all but impossible as the time taken to gain a credible estimate of its failure rate is excessive except for systems with the lower levels of safety integrity requirement. To gain confidence in the safety of a software based system both the product (the system) and the process of its development need to be assessed. Littlewood provides a good description of this topic in his paper *The need for evidence from disparate sources to evaluate software safety* [2].

This book provides guidance on the testing of safety-related software. It attempts to cover both the testing activities (*product assessment*) and the contribution that adequate performance of these activities makes to the safety case (*process assessment*) and the claimable integrity of the system.

The guidance provided is generic and represents the authors' opinions and is based upon a four year collaborative research project. The guidance is not presented as a substitute for project specific decisions on the required level and types of testing; expert advice given within the context of the particular application should always be sought.

This introductory chapter gives an overview of the testing of safety-related software and can be read in isolation. Each subsequent chapter covers a self contained topic and concludes with a list of practical guidance points.

Chapters 1, 2, 3, 5 and 9 do not place heavy demands on the reader but the remaining chapters contain more difficult concepts and may provide more challenges.

1.2 Audience

This book provides guidance on the testing of safety-related systems, and on how testing can be related to the safety case for software based systems.

The book is intended for the following types of reader:

- The system developer, who needs to be aware of the applicable methods and tools that support the testing of safety-related systems. The developer will also gain an appreciation of the role of the safety case and the principles on which it is based.

- The project manager, who needs to be aware of the important issues to be addressed in the development of safety-related systems.

- The safety assessor, who needs access to a set of guidelines on the contribution that testing makes to claims for Safety Integrity Levels (SIL).

- The post-graduate or advanced undergraduate student, who needs access to an overview of current research and development in the testing of safety-related systems.

The style and level of the book assumes that the reader has experience of the development of real time systems in engineering or industrial applications. However, although the examples used to illustrate various points are real time monitoring and control systems, the issues raised and guidance given apply equally well to high integrity systems in financial and other applications where there is the potential for large monetary loss.

It is assumed that the reader is familiar with the general issues related to the testing of software systems and the test methods and techniques commonly employed. There are many excellent books which cover this area. One good example is *Software Testing Techniques* by B Bezier [3].

1.3 Structure

The guidance provided here is directly based upon the work undertaken in the CONTESSE project, which was a DTI/EPSRC sponsored project under the Safety Critical Systems Programme.

The project title *A Critical Examination of the Contribution of Testing Using Simulated Software and System Environments to the Safety Justification of Programmable Electronic Systems* provides the main emphasis of this book. The book is divided into the following chapters:

Chapter 1, Introduction, which provides the background to the subject, describes the types of systems to which the guidance is applicable, gives a summary of the testing issues for safety-related systems and concludes with a brief discussion of current industry practices and applicable standards.

Chapter 2, Testing and the Safety Case, discusses risk assessment, the safety case and the safety lifecycle of a system. A general treatment of safety case contents is presented. The dual lifecycle which covers both system and simulated environment development is introduced.

Chapter 3, Designing for Testability, examines the common design features of a Programmable Electronic System (PES) with regard to their impact on testing.

Chapter 4, Testing of Timing Aspects, reviews the methods of validating timing requirements and also reviews scheduling strategies and the calculation of worst-case execution times.

Chapter 5, The Test Environment, discusses the requirements for test tools and reviews some of the commercially available tools.

Chapter 6, The Use of Simulators, examines the role of simulation as a test method and its use in supporting safety justification. Two models of software environment are described, the set theoretical model, and the stochastic process model.

Chapter 7, Test Adequacy, surveys test adequacy criteria and reviews their uses in the testing of safety-related software. The various ways in which adequacy criteria can be classified are briefly explained and guidelines are given for the use of test adequacy criteria.

Chapter 8, Statistical Software Testing, discusses Statistical Software Testing (SST) and its relation to test adequacy.

Chapter 9, Empirical Quantifiable Measures of Testing, presents the concepts of test cost and test regime assessment in which an empirical approach is taken to the estimation of the cost of testing and the effectiveness of the testing regime with respect to claimed integrity.

Appendix A, Summary of Advice from the Standards, contains a summary of advice on testing given in current standards and guidelines.

Bibliography, lists CONTESSE project technical reports, on which this book is based.

1.4 Applicable Systems

The book covers the testing of *safety-related* systems. These are *Programmable Electronic Systems* (PES) in safety-related applications. A *safety-related application*, in this context, can be considered as one in which a failure of the PES could result in:

- environmental pollution
- loss of, or damage to, property
- injuries to, or illness of, persons
- loss of human life or lives

to varying degrees. *Safety-critical systems* are safety-related systems that require, by the nature of the application, the highest level of integrity.

There are three basic categories of system commonly identified within the standards as having a potential impact in one or more of the above areas.

A *protection system* is a system that will set the plant (whether it be a nuclear power station, oil rig or aircraft engine) into a well-defined safe state on detection of an event that could lead to a hazard. There are two approaches to protection, shutting the system down or taking mitigating action to remove the hazard whilst keeping the plant operational. The protection system monitors the plant itself for direct indication that a hazard has occurred. This category covers those systems where there is no control system, or where the control and protection systems are separate. In the latter case the protection system may need to take action on failure of the control system as well as when the plant itself fails. It is recognised that where there is a control system then the control system itself may be safety-related, and may perform some safety-related functions. The protection system is there to cope with hazardous failure of plant or control system.

Combined control and protection systems are used where it is not possible to separate the control and monitoring functions completely. The addition of protection functions to the control system will make the control system more complex. This will increase the likelihood of faults and make the verification of the protection functions more difficult, as they will be embedded within the complex control function. In this case the design aim should be to extract as many essential safety functions as possible into a separate protection system.

Continuous control systems without protection may be used where it is necessary to maintain continuous control of a process in order to maintain safety. This is the case where there is no satisfactory safe state, and so it is not possible to produce a protection system.

1.5 Integrity Levels

Since the degree of possible damage or injury can vary markedly from one application to another it is usual to assign a *Safety Integrity Level* (SIL) to such systems. Among current safety standards four such integrity levels are commonly used (a fifth is often included simply to refer to systems with no associated safety risk). Although the precise definitions of these levels vary from one standard to another, by way of illustration we can consider those of the IEC 1508 draft standard. This standard specifies target failure measures for the four SILs for safety-related systems operating in demand mode and continuous/high demand mode as shown in Table 1-1 (note that both control and protection type safety-related systems can operate in demand or continuous/high demand modes of operation as described in IEC 1508).

Table 1-1: Safety Integrity Levels from IEC 1508 Draft Standard

	Safety Integrity Levels: Target Failure Measures	
Safety Integrity Level	Demand Mode of Operation (Probability of failure to perform its design function on demand)	Continuous/High Demand Mode of Operation (Probability of a dangerous failure per year)
4	$\geq 10^{-5}$ to $< 10^{-4}$	$\geq 10^{-5}$ to $< 10^{-4}$
3	$\geq 10^{-4}$ to $< 10^{-3}$	$\geq 10^{-4}$ to $< 10^{-3}$
2	$\geq 10^{-3}$ to $< 10^{-2}$	$\geq 10^{-3}$ to $< 10^{-2}$
1	$\geq 10^{-2}$ to $< 10^{-1}$	$\geq 10^{-2}$ to $< 10^{-1}$

This table is reproduced from [6]; © British Standards Institution.

Although this provides quantifiable target failure levels for the different categories of system, the allocation of a specific application to one of these levels may still involve a degree of subjective assessment. IEC 1508 also considers the determination of safety integrity levels using qualitative methods.

Safety-related systems are defined to be systems with SILs 1 to 4 as defined by the IEC 1508 draft standard [6]. *Safety-critical* systems are those that have the highest integrity requirement, that is SIL 4.

1.6 Typical Architectures

The selection of an architecture to meet safety requirements is based on the ability of that architecture to support the prevention or limitation of faults occurring in itself and so causing a hazardous event, or the implementation of a safety function that mitigates hazardous events produced by the plant or another system. Faults may give rise to errors in the system. Errors may cause system failures.

Random failures only affect hardware; these occur due to components wearing out and failing. Systematic failures affect all identical items, (both hardware and software) and are due to inherent design, implementation, installation and usage errors.

The major architecture design techniques recommended by the standards for preventing random and systematic failures are redundancy and diversity, respectively. These are described in more detail below. There is a lack of recommendations by the standards on alternative design techniques for systems of the higher integrity levels. However, some techniques for lower integrity systems are recommended, e.g., single channel systems with self testing and monitoring, allowing for the early detection of internal failures and failures of the output channel. This technique is recommended alone for systems of SIL 1, but may well form part of the overall design approach for constituent channels in a multi-channel system architecture for higher integrity levels.

1.6.1 Redundancy

Redundancy relies upon the principle that if a component has a certain probability of failing, then the probability of two or more identical components failing at the same time is much lower. The introduction of redundancy will therefore reduce the risk of random faults. Dual redundancy protects against the failure of one channel, whilst triple redundancy protects against two simultaneous failures thus giving higher integrity. Redundancy will reduce system level failures and so improve system reliability and hence the system's SIL.

The redundant components, or channels, of the system may be totally independent or joined by a voting component. Totally independent channels will operate without any reference to the other channels in the system. In systems where there is a voting component, allowing the cross checking of outputs, the decision to take action can be arranged either on the basis of a single channel or by agreement between two or more channels. In the case where agreement is required before action is taken, increasing the number of redundant channels may reduce the number of spurious shutdowns due to failures in monitoring or protection systems.

Redundant systems may be classified as active or passive. *Active redundancy* implies that each of the replicated channels is operating concurrently whereas systems with *passive redundancy* operate in a master-slave (or duty-standby) configuration. In the latter configuration the slave may still carry out internal self tests to maintain its own integrity and the switch-over from master to slave will probably be based on the slave detecting that the master has failed.

The category of system also influences the nature of the redundancy in the system, and therefore also affects the system architecture. For example: a duty-standby configuration is recommended for control systems where protection is not feasible. However, if this configuration is not desirable or not possible for other reasons (e.g., failure of any channel cannot be detected with confidence) then majority voting is advocated.

Dual or triple redundant systems with some comparison or voting mechanism are commonly recommended for both control and protection systems to achieve SIL 3 and 4.

1.6.2 Diversity

Replication cannot overcome systematic faults, which will affect all identical system components. The prevailing standards recommend the use of diversity for high integrity systems to overcome systematic faults. The theory behind diversity is that two components (or channels, or systems) built to carry out identical tasks, but based on different hardware, or software, or completely different technologies, will have a lower probability of containing the same systematic faults. Therefore a failure of systematic nature in one component (or channel, or system) should not cause the entire system to fail.

At different system levels, diversity may mean:

- *diverse systems* – one PES, one non-PES
- *diverse channels within the system* – different PES hardware and/or software, or one PES channel and one non-PES channel.

To achieve diversity in a single system the standards recommend that the architecture should include a PES channel and a non-PES channel. The development methods for each are independent and so the likelihood of common systematic faults should be significantly reduced. The standards also recommend that as well as making the technologies and techniques diverse, independence between different methods adopted, tools used and teams employed should be maintained.

The benefits of improved integrity through the use of diversity need to be balanced against the disadvantages. The disadvantages include:

- increase in cost or in timescales as the amount of diversity introduced increases
- increase in the maintenance and upgrade effort. The more diverse systems used, the more systems there are to maintain and upgrade
- the system may be more difficult to understand
- more complex integration path.

The diversity achieved by the use of different software development teams using different methods and tools may not be as good as one might hope. It has been suggested that the design and programming paradigm also needs to be diverse, because teams using different methods and/or tools, but still within a procedural implementation approach, will tend to make similar mistakes. What may be needed is a radically different approach, such as the use of network programming. Current work on such an approach has been investigated by Partridge [4] in the EPSRC funded Safety Critical Systems Programme project *Network Programming for Software Reliability.*

1.6.3 Guidance on Architecture and Design

Typical PES architectures have been extracted from some of the standards that offer guidance on system architecture design and are presented here by way of example. For further details of the design recommendations refer to the standards.

The HSE Guidelines

The HSE guidelines (*Guidelines for the use of PES in safety-related applications*) [5] put PES systems into perspective by comparing their architectures to those of conventional non-programmable equivalent systems, e.g., hardwired electrical or mechanical systems. No indication of SIL is given. However, the supporting text gives a qualitative feel for the safety integrity of the architecture.

For example, the guidelines state that where a conventional non-programmable single protection system would be sufficient then a suitable replacement architecture must also contain at least a single protection system. The protection system may be a single channel of programmable electronics provided that it can be shown that a failure in this single channel will not cause the system to fail in a hazardous manner. If the required integrity is such that a single channel PES cannot meet this requirement then the original system could be replaced with two PES channels. For low levels of integrity, two identical channels may be sufficient. However, if systematic faults could cause hazardous failures then the two channels should be based on diverse software and/or hardware. To give maximum diversity an alternative to two PES based channels would be to base one of the channels on non-PES technology. This is shown in Figure 1-1.

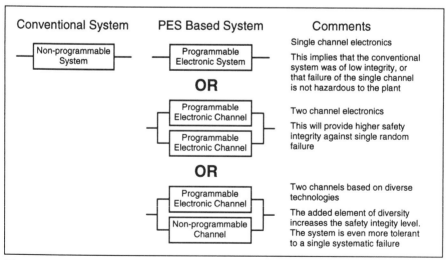

Figure 1-1: Single Channel to Dual Redundancy

This figure is reproduced from [5]; © Crown copyright.

The guidelines continue by describing example PES architectures where conventionally there were two and three protection systems. The use of more protection systems implies that the required integrity, and therefore the integrity of the replacement PES, is much higher. For an architecture requiring three conventional protection systems the guidelines state that there must be three protection systems in the new PES based architecture. The guidelines suggest that two of the protection systems may be exact duplicates but there should be at least one diverse element to make the system more tolerant to systematic failures. Again this diversity may be achieved by implementing one of the protection systems as a non-PES. This is shown in Figure 1-2.

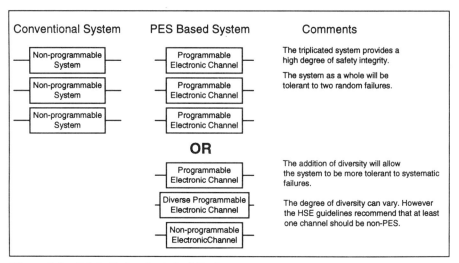

Figure 1-2: Triple Redundant Architecture

This figure is reproduced from [5]; © Crown copyright.

IEC 1508 Draft Standard

The IEC 1508 draft standard requires that the design of a safety-related system shall indicate the overall hardware and software architecture (e.g. sensors, actuators, programmable electronics, embedded software, application software etc.) that satisfies the requirements for hardware safety integrity and software safety integrity as appropriate to the application. The standard requires that the *Electrical/Electronic/Programmable Electronic Systems* (E/E/PES) architecture shall meet the requirements for fault tolerance, diagnostic coverage and proof checks as specified in Table 1-2 (safety-related protection systems) and Table 1-3 (safety-related continuous control systems).

Table 1-2 : Requirements for Hardware Safety Integrity: Components of Safety-Related Protection Systems (Including Sensors and Actuators)

SIL	Fault Requirements for type A components	Fault Requirements for type B components
1	• Safety-related undetected faults shall be detected by the proof check	• Safety-related undetected faults shall be detected by the proof check
2	• Safety-related undetected faults shall be detected by the proof check	• For components without on-line medium diagnostic coverage, the system shall be able to perform the safety function in the presence of a single fault
		• Safety-related undetected faults shall be detected by the proof check
3	• For components without on-line high diagnostic coverage, the system shall be able to perform the safety function in the presence of a single fault	• For components with on-line high diagnostic coverage, the system shall be able to perform the safety function in the presence of a single fault
	• Safety-related undetected faults shall be detected by the proof check	• For components without on-line high diagnostic coverage, the system shall be able to perform the safety function in the presence of two faults
		• Safety-related undetected faults shall be detected by the proof check

Table 1-2 : Requirements for Hardware Safety Integrity: Components of Safety-Related Protection Systems (Including Sensors and Actuators)

SIL	Fault Requirements for type A components	Fault Requirements for type B components
4	• For components with on-line high diagnostic coverage, the system shall be able to perform the safety function in the presence of two faults • For components without on-line high diagnostic coverage, the systems shall be able to perform the safety function in the presence of two faults • Safety-related undetected faults shall be detected by the proof check • Quantitative hardware analysis shall be based on worst-case assumptions	• The components shall be able to perform the safety function in the presence of two faults • Faults shall be detected with on-line high diagnostic coverage • Safety-related undetected faults shall be detected by the proof check • Quantitative hardware analysis shall be based on worst-case assumptions

This table is reproduced from [6]; © British Standards Institution.

Notice that the requirements for type A components apply if:
- the failure modes of all components are well defined
- all components can be fully tested
- good failure data from field experience exists.

The requirements for type B components apply if:
- the failure modes of all components are not well defined
- not all components can be fully tested
- no good failure data from field experience exists.

Table 1-3 : Requirements for Hardware Safety Integrity: Components of Safety-Related Control Systems (Including Sensors and Actuators)

SIL	Fault Requirements for type A components	Fault Requirements for type B components
1	• Safety-related undetected faults shall be detected by the proof check	• For components without on-line medium diagnostic coverage, the system shall be able to perform the safety function in the presence of a single fault • Safety-related undetected faults shall be detected by the proof check
2	• For components without on-line high diagnostic coverage, the system shall be able to perform the safety function in the presence of a single fault • For components without on-line high diagnostic coverage, the system shall be able to perform the safety function in the presence of two faults • Safety-related undetected faults shall be detected by the proof check	• For components with on-line high diagnostic coverage, the system shall be able to perform the safety function in the presence of a single fault and if a second fault occurs during the detection and repair of the first fault, the probability that this shall lead to an improper mode of operation of the system shall be less than the figure given for SIL 2 • For components without on-line high diagnostic coverage, the system shall be able to perform the safety function in the presence of two faults • Safety-related undetected faults shall be detected by the proof check

Table 1-3 : Requirements for Hardware Safety Integrity: Components of Safety-Related Control Systems (Including Sensors and Actuators) (Continued)

SIL	Fault Requirements for type A components	Fault Requirements for type B components
3	• For components with on-line high diagnostic coverage, the system shall be able to perform the safety function in the presence of a single fault and if a second fault occurs during the detection and repair of the first fault, the probability that this shall lead to an improper mode of operation of the SRS shall be less than the figure given for SIL 3 • For components without on-line high diagnostic coverage, the system shall be able to perform the safety function in the presence of two faults • Safety-related undetected faults shall be detected by the proof check	• The components shall be able to perform the safety function in the presence of a single fault and if a second fault occurs during the detection and repair of the first fault the safety function shall still be performed • Faults shall be detected automatically with on-line high diagnostic coverage • Safety-related undetected faults shall be detected by the proof check
4	• The components shall be able to perform the safety function in the presence of a single fault and if a second fault occurs during the detection and repair of the first fault the safety function shall still be performed • Faults shall be detected automatically with on-line high diagnostic coverage • Safety-related undetected faults shall be detected by the proof check • Quantitative hardware analysis shall be based on worst-case assumptions	• The components shall be able to perform the safety function in the presence of two faults • Faults shall be detected with on-line high diagnostic coverage • Safety-related undetected faults shall be detected by the proof check • Quantitative hardware analysis shall be based on worst-case assumptions

This table is reproduced from [6]; © British Standards Institution.

An important feature of the IEC 1508 draft standard is the support of a qualitative as well as a quantitative approach to achieving the requirements of hardware safety integrity for a safety-related system whilst still maintaining target failure measures. With respect to hardware architecture, two options are presented: the designated architecture approach and the bespoke architecture approach.

The designated architecture approach provides a qualitative means of satisfying the requirements for hardware safety integrity without the need for further analysis (providing the underlying assumptions relating to the reliability model for the architecture are satisfied). The standard recommends architectural configurations for sensors, logic systems and final elements/actuators which are appropriate for different safety integrity levels. The following designated architectures are selected in the standard:

A1   single controller with single processor and single I/O (1oo1)

A2   dual controller with dual processor (with Inter Processor Communication (IPC)) and single I/O (1oo1D) (only for PES)

A3   dual controller with dual processor and dual I/O (2oo2)

A4   dual controller with dual processor and dual I/O (1oo2)

A5   dual controller with dual processor (with IPC) and dual I/O (1oo2D) (only for PES)

A6   triple controller with triple processor (with IPC) and triple I/O (2oo3).

The selection of PES architecture for safety integrity levels 1 to 3 is shown in Table 1-4.

For SIL 4 a detailed quantitative hardware analysis is highly recommended. No designated hardware architectures are therefore recommended for SIL 4.

This approach gives a large number of potential architectures for safety-related systems since architectural configurations for sensors can be matched to architectural configurations for the logic system which in turn can be matched to architectural configurations for final elements/actuators.

The bespoke architecture approach is a more fundamental approach in which an appropriate quantified reliability model can be used to demonstrate that the target hardware safety integrity has been satisfied. Both designated and bespoke architectures must also meet specified fault count criteria.

Table 1-4 : Requirements for PES Architectures (IEC 1508)

Safety Integrity Level (SIL)	PE Logic System Configuration	Diagnostic Coverage per Channel	Off-Line Proof Test Interval (TI)	Mean Time to Spurious Trip (MTTF spurious) On-Line Repair
1	Single PE Single I/O, Ext WD	Low	1 Month	1.9 Years
		Medium	6 Months	1.7 Years
		High	48 + Months	1.6 Years
	Dual PE IPC Single I/O 1oo1D	Low	1 Month	3.5 Years
		Medium	6 Months	7.2 Years
		High	48 + Months	10 Years
	Dual PE Dual I/O NIPC 2oo2	Medium	3 Months	175 Years
		High	30 Months	290 Years
	Dual PE Dual I/O 1oo2	None	6 Months	1.4 Years
		Low	16 Months	1.0 Years
		Medium	36 Months	0.8 Years
		High	36 Months	0.8 Years
	Dual PE Dual I/O 1oo2D	None	6 Months	2.0 Years
		Low	15 Months	4.7 Years
		Medium	48 + Months	18 Years
		High	48 + Months	168 Years
	Triple PE Triple I/O IPC 2oo3	None	3 Months	8.2 Years
		Low	9 Months	8.7 Years
		Medium	36 Months	16 Years
		High	48 + Months	124 Years

Table 1-4: Requirements for PES Architectures (IEC 1508) (Continued)

Safety Integrity Level (SIL)	PE Logic System Configuration	Diagnostic Coverage per Channel	Off-Line Proof Test Interval (TI)	Mean Time to Spurious Trip (MTTF spurious) On-Line Repair
2	Single PE Single I/O Ext WD	High	6 Months	1.6 Years
	Dual PE Single I/O	High	6 Months	10 Years
	Dual PE Dual I/O 2oo2	High	3 Months	1281 Years
	Dual PE Dual I/O 1oo2	None	2 Months	1.4 Years
		Low	5 Months	1.0 Years
		Medium	18 Months	0.8 Years
		High	36 Months	0.8 Years
	Dual PE Dual I/O 1oo2D	None	2 Months	1.9 Years
		Low	4 Months	4.7 Years
		Medium	18 Months	18 Years
		High	48 + Months	168 Years
	Triple PE Triple I/O IPC 2oo3	None	1 Month	20 Years
		Low	3 Months	25 Years
		Medium	12 Months	30 Years
		High	48 + Months	168 Years
3	Dual PE Dual I/O 1oo2	High	36 + Months	0.8 Years
	Dual PE Dual I/O 1oo2D	High	48 + Months	168 Years
	Triple PE Triple I/O 2oo3	High	36 + Months	168 Years

This table is reproduced from [6]; © British Standards Institution.

DEF STAN 00-56

The DEF STAN 00-56 standard does not give specific advice on the types of architecture that may be required for the different SILs. However, it gives a scheme for decomposing systems of a required integrity level into lower level functions of different integrity levels, which when combined form a system architecture that meets the original requirement.

The standard introduces the term *combinator*, which is defined to be a physical structure or function used to combine a number of inputs into one output (some form of voting mechanism or comparator). The standard recommends that the lower level functions should be independent and where a combinator is used, it should be given a SIL equivalent to the original requirement. (This standard's definition of independent could be interpreted as equivalent to diverse.)

For example, according to the standard, a function at SIL 4 may be created from either a single SIL 4 function or two independent SIL 3 functions combined by a SIL 4 voter (as shown in Figure 1-3(a), extracted from the standard), or even a SIL 4 and an independent SIL 2 function combined by a SIL 4 voter.

For lower level functions, which are redundant but not diverse (as shown in Figure 1-3(b)), the standards recommend that each redundant function is assigned the same SIL target as the original requirement.

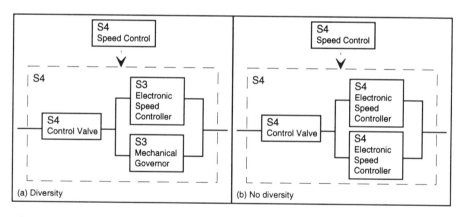

Figure 1-3: Safety Integrity Level Decomposition

This figure is reproduced from [14]; © Ministry of Defence and Directorate of Standardization.

1.7 The Safety Lifecycle and the Safety Case

To support the safety activities required for the development of safety-related systems and the preparation of a comprehensive *safety case*, DEF STAN 00-56 and the IEC 1508 draft standard both define *safety lifecycles*. Figure 1-4 shows the safety lifecycle extracted from DEF STAN 00-56. The IEC 1508 draft standard also provides a mapping of the safety lifecycle onto an overall system lifecycle showing how activities from both lifecycles correspond. The individual standards should be consulted for full details.

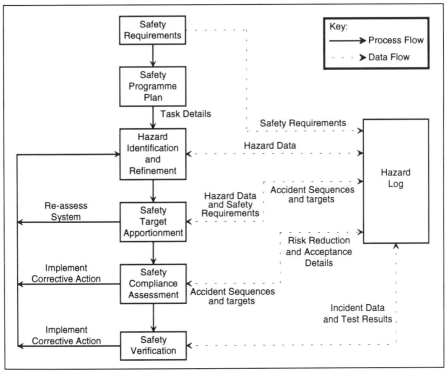

Figure 1-4: Safety Lifecycle from DEF STAN 00-56

This figure is reproduced from [14]; © Ministry of Defence and Directorate of Standardization.

Chapter 2 discusses the Safety and System Development Lifecycles, as defined in the IEC 1508 draft standard, and the link to the Safety Case.

1.8 Testing Issues across the Development Lifecycle

Issues relating testing activities within each phase of development are discussed briefly below. The requirements for tools designed to support the testing of safety-related systems are discussed in Chapter 5 which also gives a review of some commercially available tools.

1.8.1 Test Planning

To minimise commercial risk on a project, test activities must be planned in advance to determine the extent of test expenditure, which for high integrity systems may be more than half the total project costs. To assist in the development of a test plan, a structured approach to assessing test methods has been proposed and is discussed in Chapter 9.

In order to identify a development test regime that is commensurate with the required integrity of a system it is necessary to make a qualitative assessment of alternatives, and this is effectively the approach inherent in all of the existing and emerging standards. However, two main problems can be identified with the approach taken by many of the standards:

- each SIL covers an order of magnitude in reliability, and
- there is little notion of test adequacy and the system developer is left to make a judgement as to what is required.

In an attempt to overcome these problems the *test regime assessment model*, described in Chapter 9 can be used both to plan the testing programme and also to allow continuous qualitative (but structured and auditable) assessment of systems. The model also takes cognisance of issues such as:

- the degree of correspondence between simulated environments and the final system environment
- the coverage provided by tests, particularly with regard to testing of the system's response to the unsafe external and internal states identified during hazard analyses
- changes to system design between tests
- human experience, competence and independence.

1.8.2 Requirements Phase

Verification, which occurs at the end of each life cycle phase before moving on to the next, is often described as ensuring that we are building the product correctly, while validation, which generally occurs at the end of the life cycle, is described as ensuring that the correct product has been built. It is of course much too late to discover that we have the wrong product after the very expensive task of building it has been completed. However, animation of the requirements for customer feedback is a validation activity.

One of the fundamental questions to be answered is: 'Are the requirements correct?'. The answer is usually provided by a review. Review methods rely on the interpretation of specifications for behaviour. Introduction of explicit behaviour modelling conventions removes ambiguities, furthermore if these conventions are supported by an executable system, then the requirements can be animated and dynamic test techniques can be invoked. Thus specification animation that allows interactive testing of the system by the end user, as well as a systematic approach to testing, will offer important benefits. It is often important to carry out checks on the likely performance of the system to be developed during the requirements phase. A performance model of a candidate architecture can be built and run against various plant operational scenarios.

Once the requirements have been identified, it is important that compliance with the requirements is maintained throughout subsequent development and test activities. In order to do this, traceability links are maintained which cross reference between items produced in development of the system and the requirements, identifying which requirement or requirements each item satisfies. The fundamental relationship between requirements and test specifications cannot be over emphasised. Particular emphasis should be placed on demonstration of functions deemed to provide safety functions. This means that requirements tracing and demonstration of safety functions should be segregated.

1.8.3 Design Phase

Data flow based design, data structure based design and object oriented design are three design methods that are currently popular. These methods have not been developed with application to safety-related systems design in mind and so there are some issues to be addressed. With data flow based design the translation from analysis (data flow diagrams) to design (structure diagrams) is an ill defined process and requires particular attention to maintenance of traceability links. Object oriented techniques have an implied concurrency in the design that may not be realisable within the constraints of accepted coding practices for safety-related systems.

The software design should be checked to ensure that it is compatible with the test methods laid out in the test plan and consideration given to the introduction of interfaces specifically for test purposes. For safety-related systems greater emphasis must be placed on compliance with the test plan because compliance will form part of the safety justification for the system.

The hardware architecture should be considered in relation to the test plan and where possible the system should be decomposed into dedicated processors providing specific functions. The hardware interfaces introduced may afford a non-intrusive means of monitoring system performance during tests.

When choosing commercially available microprocessor based systems, the mode of operation of the system should be considered. Systems with a well-understood time deterministic mode of operation should be preferred. Systems that employ general purpose multi-tasking operating systems should be treated with caution as they permit the introduction of undesirable behaviour. Multi-tasking systems may make the system response non-deterministic and make the dynamic testing more problematic due to variations in the observed behaviour between repeated runs of the same tests. More sophisticated analysis of results will be required to explain these variances.

1.8.4 Implementation Phase

Implementation languages that are well structured and offer clear interfaces between modules should be preferred as this will facilitate module based testing. Advice on avoidance of unsafe language constructs should be sought and implemented via enforcement of coding standards. In some cases tool support may be available to support demonstration of compliance with the coding standards.

Static Testing

Static testing techniques can range from a code inspection by the authors through to full semantic analysis and assertion checking using static analysis support tools. Intermediate levels include data and control flow analysis. At the 'high' end, static analysis is a costly exercise. For some implementation languages there is no tool support and therefore only the lower or intermediate levels of analysis can be carried out using 'in-house' developed software. Code inspection should be carried out in all cases. A 'risk based' approach should be adopted to make cost effective decisions on the application of the more rigorous techniques.

Dynamic Testing

Dynamic testing should be based on a three pronged approach prioritised in the following order (high to low):

1. functional testing
2. coverage based testing
3. random input testing.

Functional testing must always be carried out to demonstrate that the system performs the functions requested of it. *Coverage based testing* should aim to explore the system's behaviour around its boundary conditions and further coverage criteria, e.g. statement, branch, condition, multiple condition, basis path, etc. may be employed.

Random input testing is a technique that is encouraged by some certification authorities. It can provide the following benefits:

- gaining confidence in a system's functional correctness
- examining a system's response to adverse operating conditions
- predicting a system's reliability for quantitative integrity analysis.

The testing is achieved using a simple simulation of the system and its environment, which is exposed to semi-random test trajectories.

The test methods chosen should be applied at appropriate levels of the system.

Table 1-5 : Applicability of Test Methods to Levels of Testing

Method	Module Level	System Level
Static Analysis	Yes	No
Code Inspection	Yes	No
Functional	Yes	Yes
Coverage Based	Yes	No
Random Input	No	Yes

Test Results Analysis

When analysing results, the comparison of predicted response and actual response of a system will be complicated by the slight variances that will occur. The comparison technique employed must have some means of accommodating these small but generally acceptable variances otherwise a large number of test 'failures' must be explained. Introduction of analogue values in the system inputs or responses will add to the difficulties in analysing whether the response is acceptable.

For tests specified in terms of static pre-conditions and post-conditions the problems of comparing actual and predicted responses may be reduced as they will generally be expressed in terms of static conditions.

1.9   Tool Support

Chapter 5 contains a discussion on the requirements for automated tools to support the testing of safety-related systems and a review of some commercially available tools. In Chapter 6 a particular class of such tools, the Plant Simulator to which the safety-related system interfaces, is discussed separately.

The summary requirements for a set of tools that will support the testing of safety-related systems are listed below.

- A simulation tool that permits functional requirements to be modelled and executed and allows for systematic testing to be applied much earlier in the development lifecycle.
- A tool that maintains links between requirements, test cases and system elements to be tested, and can report on these links and the status of each of the related objects.
- A tool that supports calculation of test regime compatibility with required integrity.
- A complexity analysis tool for control of development of code.
- A code coverage analysis tool based on actual execution of code.

- A test case management system to create and monitor and report on test case execution.
- A simulation tool for stochastic environmental modelling. Such simulation models can be used to 'close the loop' to provide a system under test with feedback of controlled system response.
- A system test harness, possibly configurable, to stimulate and record system responses.
- A tool that compares and/or analyses results from a system simulation and from the actual system itself.

1.10 Current Industrial Practice

The regulatory bodies for safety in a selection of industries are listed in Table 1-6.

Table 1-6 : Industries and the Safety Regulatory Bodies

Industry	Regulator / Enforcement Agency
Chemical Processing	Health and Safety Executive
Gas Industry	Health and Safety Executive
Offshore Installations	Health and Safety Executive
Process Industries	Health and Safety Executive
Nuclear Power Generation	Health and Safety Executive, HM Nuclear Installations Inspectorate
Civil Aviation	Civil Aviation Authority
Commercial Shipping	DoT Marine Directorate, International Maritime Organisation, Certification/Classification Societies
London Underground	Department of Transport, HM Railway Inspectorate
Railway Signalling	Department of Transport, HM Railway Inspectorate
Defence	Ministry of Defence
Mining	HM Inspectorate of Mines and Mineral Workings

The approach to safety of a selection of the industries is reviewed to assess current industrial practice with regard to testing of software systems. The aim of this review is to determine the importance typically placed by industry on different test methods when preparing a software safety case and how those test methods vary with the level of safety integrity that is to be achieved. An analysis of the results of an industry review is provided in the following paragraph.

1.10.1 Approach to Software Safety Taken by Industry

The approach to the safety assessment of software within industry varies considerably. Some industries are extremely rigorous in their approach and require that their software suppliers follow recognised safety standards or guidelines. Such standards and/or guidelines include, for example, the HSE *Guidelines for the use of PES in safety-related applications* [5] and the International Electrotechnical Committee guidelines on *Software for Computers in the Application of Industrial Safety-Related Systems* [6]. The systems supplied are assessed against these standards or guidelines and in some cases, for example the nuclear industry, also against the industry's own safety assessment principles. Other industries are less rigorous having only recently mandated the presentation of safety cases and as yet provide little formal guidance to software suppliers on software safety issues.

All industries recognise that software safety is an important issue that cannot be addressed by traditional methods more suited to the analysis of the random failures in hardware systems. They also recognise that as yet there is no widely accepted method of software safety analysis, however the reaction to this uncertainty varies. Typical reactions include:

- Avoid the problem – don't use software.
- Use only bespoke software that has a 'known' high level of quality.
- Keep it simple – it is believed that if it is simple enough then it can be adequately verified.
- Use trusted suppliers experienced in safety-related system supply.
- Use recognised software safety standards.
- Define safety assessment principles.

Many industries, of course, adopt more than one of the above approaches.

1.10.2 Specific Approach to Software Safety by Industries

The following paragraphs provide an outline of the approach to software safety taken by:

- the Nuclear Power Generation Industry
- the Railway Industry
- the Civil Aviation Authority
- the Process Industry.

Nuclear Power Generation

The regulator of civil nuclear power installations within the UK is the Nuclear Installations Inspectorate (NII). The NII recognises that the quantitative assessment of software based systems is not currently achievable and so safety assessment for these systems is based on qualitative aspects. For the production and operation of software based nuclear protection systems, the NII require that very high quality and technical standards are employed. Minimum quality standards, for example IEC 880 [7], are to be adhered to for all software protection systems development. Technical standards are not mandated, but must be relevant to the level of reliability required and must be consistent with currently accepted industry standards for safety-related software development.

The NII requires that comprehensive supplier verification and validation activities are undertaken by the system supplier throughout the software development process. This is to be followed by an entirely independent assessment of the system, encompassing design assessment and full system test. Some guidance is provided by the NII on the level of coverage that is to be achieved during testing. The Verification and Validation (V & V) activities undertaken by both the system supplier and independent assessor include static and dynamic testing.

To verify that the practices adopted during development of a software system actually achieve the required level of reliability, the NII recommends that a probabilistic estimate of the software reliability be 'computed'. This value is compared with that implied by the development methods applied. If the computed reliability value is less than the implied value, then it is to be assumed that the applied methods have been ineffective. The implied value is to be revised accordingly. To date this recommendation is purely a recommendation 'in principle' and has yet to be achieved in practice. No recommendations are provided by the NII on possible approaches to the computation of software reliability.

It is not apparent that any more importance is placed on any one aspect of software development than another. Quality control and assurance, design practice, documentation and testing are all important to the assessment of the system and must be undertaken to standards generally accepted as suitable for safety-related systems. If any one aspect of development does not meet the required standard, the NII are unlikely to accept the system.

It is likely that techniques such as dynamic testing that support the numeric reliability estimation of software, will become increasingly important given the NII recommendation that a probabilistic estimate be computed.

Civil Aviation Authority

Within the field of civil aviation there is no specific standard that must be followed for the development of software. Instead the Civil Aviation Authority (CAA), the UK regulator, follows internationally accepted guidelines for software development for airborne systems or equipment (RTCA/DO-178B) [8]. These guidelines are non-prescriptive and contain advice on the safety objectives that should be met when producing software. The RTCA guidelines' definition of integrity levels is different from that found in the IEC 1508 draft standard, however, the concept of levels of integrity is common to both. The objectives stated in the guidelines may vary depending on the safety criticality of the software. The guidelines include, for example, definitions of integrity levels, guidance on design methods, programming languages, levels of documentation required, test methods and test coverage. The means of achieving the stated objectives are left entirely to the supplier. For land based systems no equivalent guidelines are in place, but the CAA is starting to apply the recommendations of the IEC 1508 draft standard [6].

The CAA assesses the safety case for any safety-critical airborne system or equipment containing software against their industry guidelines. If these guidelines have not been followed by the supplier, then the acceptability of any development standards and/or methods that have been employed will be judged against them.

The guidelines indicate that both static and dynamic testing are important V & V techniques. Guidance is provided on test coverage criteria to be met for systems of differing levels of safety criticality and also the level of independence required of the tester. Target environment and processor simulation are accepted as adequate validation techniques although high fidelity simulations are required. The guidelines also indicate that exhaustive input testing is an acceptable alternative to the software verification process. This is only the case however where the system under test is 'simple and isolated, such that the entire input and output space can be bounded' [8].

Railway industry

The railway industry within the UK is regulated by HM Railway Inspectorate. This body along with the British Railways Board and its research department and a number of other interested parties have developed an industry specific standard, RIA 23 [9], for the procurement and supply of programmable electronic systems for use in railway signalling applications. This standard is based on an early version of the IEC 1508 software standard [6] and defines how this standard is to be used in railway signalling applications. RIA 23 is currently a consultative document and has not yet been released for formal use. Although it has been developed initially for signalling applications, it is intended that it may form the basis of a standard for a wider range of railway applications, including traction and rolling stock systems. RIA 23 is likely to be superseded by the draft CENELEC standard, Railway Applications: Software for Railway Control and Protection Systems [10].

The RIA 23 standard [9] defines the lifecycle model to be adopted for software development, specifying the activities to be carried out in each development phase along with required inputs and outputs. In common with many standards, the emphasis is as much on the quality of the entire development process as any quantitative guidance on particular methods. Rather unusually however the standard does provide lists of recommended methods applicable to each lifecycle phase and grades their importance, albeit subjectively, to the level of safety integrity that is required. This stems from the fact that the standard is based on the IEC standard that adopts the same approach.

For the testing phases of development a variety of both static and dynamic test methods are recommended by RIA 23. This recommendation varies from 'useful' to 'mandatory' depending on the perceived value of a particular test method and also the integrity level. Simulation is considered a useful technique for all levels, but it is noted that the development of test software should be carefully controlled as this software contributes to the totality of the software under test and may itself pose problems in validation. It is not apparent that dynamic testing is regarded as any more or less important than static testing. Some techniques from both methods are strongly recommended for all levels and in some cases even mandatory. For full details refer directly to the standard. Some guidance is also provided on test coverage but this is not classified by integrity level as such classification is not judged to be possible.

RIA 23 requires that software verification and validation is carried out by independent teams. The independent team are to audit and review the development verification and testing activities and also to perform their own independent analyses on the software and its documentation.

Process industries

There is no governing standard or regulatory authority in the UK for the process industries, in the same sense as there is for the nuclear or civil aviation industries. These roles have been assumed in the past by the Health and Safety Executive (HSE) PES guidelines and the HSE itself. The HSE PES guidelines were issued some years ago and will effectively be superseded by the documents from the IEC 1508 standard.

The development of the International Electrotechnical Commission (IEC) 1508 draft standard as a generic standard for hardware and software in safety-related systems is achieving a consensus across industries and countries. It is drawing together the work of bodies such as DIN in Germany and the Instrument Society of America (ISA).

However, the IEC document is still in draft form. In the interim period the PES guidelines are considered to be too general, hence a number of guideline documents are emerging to fill the gap from sector specific bodies, such as the UK Offshore Operators Association (UKOOA), the Engineering Equipment and Material Users Association (EEMUA), the Institute of Gas Engineers and individual companies such as BP.

The danger in this is that each guideline document could adopt a different approach and by the time the IEC standard is issued so many installations will be in place that have not been developed according to the principles defined by the standard that it will not be adopted. This danger seems to have been averted as the working groups for IEC 1508 have succeeded in achieving a high level of visibility and reflecting a general consensus. Industry specific and country specific standards and guidelines are being developed taking into account the IEC work.

The working groups for IEC 1508 have included one group working on the generic standard and one working on the sector specific standards such as medical, transportation and the process industries. Within the sector specific part, Task Group C has been established for the process industries. The ISA in America has formulated its own standard (ISA SP84 [11]) for the use of PESs in safety-related applications in the process industries and has submitted it for consideration for adoption by the IEC. Task group C will harmonise the ISA document with the generic standard of the IEC.

The ISA standard was given much support in the USA when the Occupational Safety and Hazards Administration (OSHA), a cabinet branch of the Government, based cases of criminal liability on non-conformance to ISA SP84. This was very effective in generating a desire, previously lacking on the part of system vendors and operators, to contribute to its content.

In Europe, DIN 0801 [12] addresses hardware and software safety in microcomputer control systems and is evolving to harmonise with the IEC standard. The German approval bodies, the regional TÜV-N (Technical Inspection Association of Northern Germany), use the DIN standard, amongst others, in the assessment of systems.

The DIN standard is prescriptive, specifying the mechanisms for implementation whereas the IEC approach is to state the objectives and principles to be achieved. Although prescriptive, the DIN standard requires some interpretation. This is done by the regional TÜVs, when assessing systems for approval. Many companies, particularly in the offshore oil and gas sector, demand TÜV approval of equipment used in emergency shutdown and fire and gas systems.

1.10.3 Industry Use of Standards and Safety Assessment Policies

A number of industries, (listed in Table 1-7), have industry specific guidelines and standards against which safety-related software is developed and assessed. Only one industry, Defence, mandates the use of an industry standard. In all other industries the use of standards and/or guidelines, although strongly encouraged, is purely advisory.

Table 1-7 : Industry Specific Standards and Guidelines

Industry	Standard
Defence	DEF STAN 00-55 [13] and 00-56 [14]
Nuclear Power	IEC 880 [7]
Civil Aviation (airborne systems only)	RTCA–178B [8]
Gas	IGE/SR/15 [15]
Railway	RIA 23 [9] CENELEC [10]

Typically some justification will be required, and the agreement of the regulator sought, if an alternative standard and/or guideline is followed. This book does not attempt to identify or comment on all standards relevant to the development and testing of safety-related software. An excellent comprehensive review of these standards can be found in *Guide to Software Engineering Standards and Specifications* by J Magee and L Tripp [16]. This is the most detailed reference available and provides a directory and an overview of more than 300 standards and specifications written in the English language. It is intended to be updated every 18 months. The entries are categorised by classifications such as 'safety.'

Industries that have not developed their own standards or guidelines tend either to avoid using software at all in a situation where a safety hazard may result, or to require only that an *appropriate* safety standard is applied – leaving the choice of standard to the supplier. In the latter case a software safety assessment is still normally carried out on all supplied software systems, although the assessors are usually not provided with formal guidance on the interpretation and/or applicability of available safety standards to the particular industry and/or application.

Lack of direct guidance on the assessment of software does not necessarily mean that industries are not conducting adequate safety assessments of software, but it does mean that assessments are very much dependent on the experience and knowledge of the individual assessor. In industries with very little experience in software safety this is likely to be more of a problem than in industries with a well-established record of safety awareness. London Underground for example, although having neither a formal safety assessment process for software nor mandating any particular safety standard, has a very experienced assessment team, some of whom are serving on IEE committees considering software safety issues.

The BSI is currently producing a new standard, BS5760 part 8, *Assessment of the Reliability of Systems Containing Software* [17]. This is currently in draft form (*Draft for Development 198*) but should be published soon.

1.11 The Significance Placed upon Testing by Standards and Guidelines

Current standards and guidelines for the development of software used in safety systems, provide guidance on the entire process of system development and documentation. They require that a process is followed which is conducive to the production of a safe, and high quality, product. The product is to be continuously assessed throughout development and operation to ensure continued compliance with its safety requirements. This assessment is, however, purely a subjective judgement of the development methods employed, against the requirements or recommendations of the particular standard or guideline and also currently accepted good practice.

There is no standard or guideline that provides guidance on techniques aimed at providing a quantitative measure of software safety and few that even provide guidance on how the level of safety integrity achieved may be justifiably implied by the development process methods. The standards and/or guidelines simply state the safety objectives to be met but do not indicate how these objectives may be achieved. The exception to this is IEC 1508 draft standards (and RIA 23 [9]) which give recommendations, varying depending on the level of required safety integrity, on suitable development techniques and/or measures to be employed.

All the standards and guidelines reviewed discuss the process of testing safety-related and safety-critical software. This discussion however normally revolves around the process of testing (phases involved, plans and specifications to be produced, level of independence required, records to be kept, etc.) rather than specific test methods and/or techniques to be employed.

The level of advice given on test methods and techniques varies considerably. In some cases it is extremely limited whereas in others it is fairly comprehensive. Appendix A provides a summary of the testing advice provided. The aim of the appendix is to give an overview of the level of test advice available, rather than to repeat that advice in full. If specific test techniques and/or test coverage criteria are recommended then this is indicated but details are not provided. The reader should refer to the source documents for more information.

1.12 Guidance

> **Summary Guidance from the Standards**
>
> - The rigour with which testing must be carried out increases with safety integrity.
> - The level of testing independence required during different test phases depends on the level of safety integrity.
> - Formal (mathematical) methods and proofs although mandated or at least strongly encouraged by some, are still regarded by many as too immature and costly for general use.
> - Both static analysis and testing, and dynamic testing are regarded as important verification and validation techniques and are considered essential for high integrity systems.
> - Resource modelling and performance testing are considered essential for high integrity systems.
> - Both black box and white box testing are required at all levels of integrity.
> - Full functional testing is required for all levels of integrity.
> - The test coverage to be achieved during testing must be prespecified, justified as adequate, and its achievement must be demonstrated.
> - The use of simulators, including environment simulators, is considered useful (in some cases unavoidable) during testing but with some reservations:
>   - The simulator may be complex and pose its own validation problems.
>   - All test software, including simulators, contribute to the totality of software developed and must be carefully controlled.
>   - To be valid, simulations must be shown to be accurate representations of that which they are simulating (i.e. high fidelity simulation).
> - Use of automatic test tools is encouraged both during testing and for the derivation of test cases.
> - Statistically valid random testing covering the normal operational environment of the system, abnormal conditions, etc. resulting in a *quantitative* reliability value, is increasingly being considered essential for high integrity systems, although not everyone as yet subscribes to this view.
> - The use of certified or validated tools is encouraged.

Chapter 2
Testing and the Safety Case

This chapter discusses risk assessment, the software safety case and the safety lifecycle of a system. Testing, and its contribution to safety justification, is set within the context of an overall safety case and the safety lifecycle.

A safety case provides a framework within which the safety aspects of a system may be detailed and assessed. A general treatment of safety case contents is presented. It is proposed that such safety case contents can be derived from a generic format, and that the contents and lifecycle stages in any particular instance will be industry and project specific.

The dual lifecycle which covers both system and simulated environment development is introduced. The link between the dual lifecycle, safety and test activities is discussed.

2.1 Introduction

The uncertainty associated with both the adequacy of requirements specifications and with our ability to translate these requirements into a final product means that we ultimately rely upon verification and validation (V&V), including testing, of the design through its various levels from initial concept to final implementation.

This is particularly the case with the final implementation, where we are no longer checking a specification, but are concerned with the hardware and software (code) within which the specification is finally realised.

In terms of safety, we seek to perform various analyses and assessments to give us a measure of confidence that hazardous situations which can be linked to the system being developed have been identified and that measures are applied to reduce the risk of occurrence of the hazards to a tolerable level.

The achievement of the required Safety Integrity Level (SIL), as determined by risk analysis, must be demonstrated to support safety claims for the system. Testing, and in particular safety testing, is one of the principal means of demonstrating that the claimed software safety integrity is in accordance with the safety requirements.

Safety, and the manifestation of faults as hazards, are system level issues. The software itself may not be unsafe or hazardous but, within a system context, it can produce the means by which a hazard may be initiated.

It is the recognition that such cause-effect relationships exist between components or functional areas of software that results in the identification of safety-related items of a system. These safety-related items may be a possible cause of a hazard or have a role in preventing a hazard (i.e. they perform a safety function).

Where such safety-related items are implemented in software, the achievement and demonstration of software safety integrity requires that conventional software V&V be extended to incorporate a system safety perspective.

As part of safety V&V activities and overall testing we seek to ensure that all identified hazards have been eliminated or that the risk has been reduced to an acceptable level. Tests performed to demonstrate the adequacy of the software design to perform safety functions or to avoid faults that lead to hazards are identified as safety tests.

2.2 Safety and Risk Assessment

Within the analysis performed during a risk assessment we are generally concerned with examining both the system design and the robustness of the system in terms of deviations from the design. Thus from a safety perspective, we wish to assure ourselves that the system will be safe both in normal operation and in the presence of faults.

The general stages of risk assessment are embodied in safety standards and guidelines (such as the HSE PES guidelines [5]), and form the basis of design and assessment procedures. These procedures may be summarised as follows:

Step 1: Analyse the hazards:
- a identify the potential hazards
- b evaluate the events leading to these hazards.

Step 2: Identify the safety-related items. These are items which can affect the safety integrity of the system and whose failure may be implicated in the events leading to the hazards identified above.

Step 3: Decide on the required level of safety integrity for the safety-related items.

Step 4: Design the safety-related items using the recommended techniques appropriate to the specific application and the required Safety Integrity Level.

Step 5: Carry out a safety integrity analysis to assess the claimable level of safety integrity.

Step 6: Ensure, from the safety integrity analysis, that the required level of safety integrity, has been achieved, or that appropriate corrective action is taken.

It is within the risk assessment strategy, outlined above, that the cause-effect relationships which include the safety-related items are identified. Where design solutions to prevent, control or mitigate hazards are implemented in software, the cause-effect modelling of the software safety functionality will form the basis of software safety testing.

The requirements for testing the safety aspects of the software are generated at step 4 with the results of such testing, which will form part of the overall safety assessment and will support the safety case for the system, coming out of steps 5 and 6.

2.3 Hazard Analysis

There are several techniques that are used to identify hazards and assess risk. A summary of Hazard Analysis (HA) techniques commonly used in risk assessment is given in Table 2-1 at the end of this chapter.

In the development of high integrity software, significant advances have been achieved in eliminating software implementation faults. However, numerous studies together with industry experience indicate that the most significant sources of software initiated failures are faults introduced at a requirements and specification level, and the interaction of software with its hardware and system environment.

As such, the notion of a software fault can frequently only be realised within a system context. This is particularly the case with safety, where hazards manifest themselves in a physical manner at a system level. Consequently any fault analysis of a design depends upon the ability of the design process to incorporate the system level perspective. An essential aspect of a risk based approach within software development is the imposition of design and assessment criteria from a system safety perspective.

Analysis of failure mechanisms within hazard analysis may be extended into software from a system perspective to provide confidence on the absence of specification based faults and assurance of correctness. This requires an integration of system safety and software techniques whereby hazard analysis techniques are applied in conjunction with conventional software analysis techniques.

2.3.1 Hazard Analysis and The System Lifecycle

As a project progresses, design decisions lead to the generation of increasing levels of design detail. Figure 2-1 illustrates the role of risk assessment techniques during the system lifecycle, and portrays the iterative processes which accompany design decisions and the generation and identification of safety-

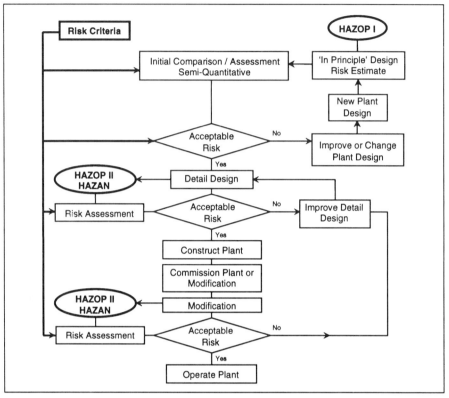

Figure 2-1: Typical Steps in Risk Acceptance Procedure

related items during the project lifecycle. The initial comparison/assessment (of risk) is carried out on the system design.

Safety-related items will be those parts of the design that will cause disproportionately more damage, disruption, loss of function etc. to the system than other parts. Identification of these items results from the realisation that a cause-effect relationship exists between two or more items. The notion of cause and effect is fundamental to hazard analysis methodology. There is no single technique that is capable of identifying safety-related items in isolation and identification also requires continual interaction with the main project management activities.

The techniques listed in Table 2-1 are typical of the hazard analysis techniques which will be applied to the software and system in accordance with the level of design detail available and the requirements of risk analysis.

In general, the demonstration of component and/or system integrity characteristics may take evidence from the processes used for development and from the product itself. Demonstration of integrity from process evidence is based upon the rigorous application of appropriate methods and techniques. Product based evidence may include historical or statistical evidence, analysis of physical characteristics or test results. For software, testing is one of the primary sources of product evidence. Other product evidence includes the results of static analysis, code inspections, and document reviews. It is the systematic approach to defining the requirements for software safety testing which is of prime importance. In particular, hazard analysis provides the means for examining software and system design at the specification level and the justification that tests on the implemented software are sufficient to support the safety case.

2.3.2 Software Hazard Analysis

Until recently, hazard analysis of computer based or Programmable Electronic Systems (PES) has seldom been carried out inside the boundary of the computer. Typically each computer and its software has been regarded as a single unit with no attempt made to analyse its various component parts. Current standards and guidelines for safety-related systems now suggest that if sufficient functional independence can be shown, individual subsystems within the computer system may be separately analysed and allocated different SILs. Guidance is given in the IEC 1508 draft standard on the degree of independence required to allow the allocation of separate SILs. To allow separate software components to be allocated different SILs, there must be clear evidence of component independence, in terms of both functionality and method of implementation. In reality, this independence is difficult to achieve.

Current standards and guidelines for the development of the software element of safety-related systems are now encouraging the application of hazard analysis to software. The aim is to identify the most critical areas of the software, perhaps even down to module level. If this can be achieved, then effort can be focused on these critical areas and fault tolerant design techniques, more rigorous development methods and controls, and more extensive verification and validation can be implemented. In this way greater confidence can be placed in the software performing its safety functions or that the software will not cause a hazard by its own actions.

What the Standards Say

Although standards and guidelines encourage the use of software hazard analysis, the assistance they provide on the topic is very limited. Some of the objectives for the analysis included in the standards/guidelines are:

- to identify critical system modules and program sections
- to verify that software required to handle failure modes identified by systems/subsystems hazard analysis does so effectively
- to allow more rigorous methods and controls to be selected and applied to areas of software which are most critical to the safety of the system
- to identify and evaluate safety hazards associated with the software, with the aim of either eliminating them or assisting in the reduction of associated risks

- to identify failure modes that can lead to an unsafe state and make recommendations for changes
- to determine the sequence of inputs which could lead to the software causing an unsafe state and make recommendations for changes.

Approaches suggested include Failure Modes and Effects Analysis (FMEA) and Fault Tree Analysis (FTA). The HAZOP technique has also been recommended (DEF STANs 55 and 56 [13], [14]) and has been the subject of recent interest (CONTESSE report [18]) ([Interim Defence Standard [19]). For programmable electronics and its software the HSE Guidelines [5] suggest that a functional approach to FMEA is adopted as opposed to the more conventional approach. The Canadian Ontario Hydro standard [20] requires that a code hazard analysis be undertaken and provides a list of issues which should be considered during such an analysis.

Threats to Consider During Software Hazard Analysis

Current software safety standards provide little advice regarding threats which ought to be considered when undertaking a software hazard analysis. In an attempt to fill this gap examples of some general threats that will apply to the design of any software based systems are identified and discussed below. Analysis of these and other threats will assist in identifying critical software modules and unsafe failure modes. This knowledge may then be used to change the design to reduce the risk of a hazard occurring or to direct the more rigorous development and fault tolerant design techniques to the most critical areas of software.

Environmental and Operating Conditions   The specific environment or operating conditions in which a software component is executed may affect the operation of that component in an unexpected way. Threats associated with the expected environment / conditions should be considered.

An example scenario for real time systems is the case where a software function disables interrupts and thus prevents their timely servicing. If the design of the system is such that a safety-related function is interrupt driven, or relies on some other event which is interrupt driven, then it may fail to carry out the critical function correctly or fail to perform it at all. Such a situation may be very difficult to detect during testing as it may rely on a critical timing of events which seldom occurs. A functional hazard analysis which considers the threat 'function fails to be invoked' would, however, be likely to detect the problem.

Logic Control (Real-Time Executive)   For asynchronous systems the logic control, or order of processing, of components can be very difficult to predict, particularly when the aim is to identify potential failure conditions. To ensure hazardous failure modes relating to an unanticipated processing order are not missed, it is suggested that a software hazard analysis should make no assumptions about the executive control imposed on the logic (Leveson and Harvey [21]). This should allow hazards resulting from unexpected control to be identified and protected against.

System Function Calls Computer operating systems normally provide libraries of standard functions which can be called from software programs to carry out routine tasks (e.g. disk and file access). These functions are designed to be both time and resource efficient by taking advantage of the facilities inherent in the computer architecture. From a safety point of view there is a drawback to the use of such functions as their internal design may be unknown and hence their failure modes and effects difficult to analyse. To allow a comprehensive software hazard analysis to be undertaken for a system which makes use of such functions, it is essential to have a detailed understanding of their operation. If this information cannot be obtained it may be necessary for safety-related systems to avoid their use and implement such functions within the application itself.

System Resources The precise usage at all times of system resources (disk space, memory, etc.) within a software system is not easy to predict. Failure to gain access to required resources may cause a software function to fail, possibly in a very unexpected manner. It may even cause the failure of a supposedly 'independent' function.

A software hazard analysis should consider system resource usage and identify where failure to obtain a resource will have a hazardous consequence. This should be relatively straightforward for those resources which are explicitly requested (e.g. an I/O channel), but may be much more difficult, or even impossible, for those which are implicit (e.g. access to stack). A hazard analysis should at least attempt to identify what resources are used implicitly and ensure that some safety function is defined which, for example, monitors their usage and raises an alarm when the resource levels fall below some 'safe' limit.

Timing For many systems the exact timing of issuing commands or carrying out computations may be essential to the safe operation of the system. Further, the maximum time taken to complete an activity may be critical. A software hazard analysis should identify which timing issues are critical to safety so that design reviews and tests can ensure that these timings are achieved or compensating action taken if this is not possible.

Software Design It is anticipated that some software design notations or representations may be more amenable to hazard analysis than others (e.g. a modular structure may assist functional analysis). Similarly the choice of design representation may facilitate or hinder the introduction of safety features to protect against identified hazardous failure modes.

It may be argued that hazard analysis techniques offer no advantage at certain levels of software design detail, and that existing software modelling and analysis techniques are more appropriate and effective at the level of assessment required.

For example, at the more detailed specification levels, static analysis may be more appropriate, or hazard analysis may be complementary. Similarly, the information generated by hazard analysis will feed into conventional software testing and simulation activities, the results of which feed data back to the failure models.

2.3.3 Hazard Analysis and Testing

The subject of test case design can be considered to be the crux of software testing. In general, tool support for testing has been directed at code level testing activities. Tool support for test case generation using a specification as input has, however, remained problematic, particularly where the descriptive medium for the specification is imprecise (e.g. natural language).

The generation of test cases and test data will be dependent upon some model of the system to be tested and the choice of test methods to achieve the required testing. Rigorous testing can be achieved by using black box testing and white box testing in a complementary manner. Myers [22], for example, recommends that test cases be developed using black box methods; examining functionality; with supplementary test cases developed as necessary using white box methods; examining the software logic.

Hazard analysis supports black box modelling of systems with supplementary white box modelling where necessary, in that they may be applied to the modelling of cause-effect relationships at an arbitrary level of detail.

Through modelling provided by hazard analysis, sequences of demands are executed which reproduce hazardous failure scenarios as presented in the modelling of failure mechanisms. Where we have asserted that software provides a contribution to the safety measures outlined earlier, the tests and simulation must demonstrate that the assertions are justifiable, and that the system is robust under fault conditions. The cause-effect descriptions of failure mechanisms from hazard analysis may be used to describe such demand scenarios. It is in this way that hazard analysis may contribute (and for safety-related systems should contribute) to software testing via simulation This topic is discussed more fully in Section 8.5.3.

In general, hazard analysis can offer benefits at two levels when applied to the PES and software aspects of safety-related systems.

Firstly, at the conventional environment level, hazard analysis provides a means of describing the intended design and fault conditions under which the software is expected to function. This information contributes to software design decisions as well as the generation of requirements for testing against environment scenarios.

Secondly, hazard analysis may be extended into the PES and software. In this case, it helps establish a clear understanding of the design intent, and will be particularly useful in identifying robustness and failure related characteristics not apparent from conventional analyses. The structured modelling of PES and software functionality will be carried out as an aspect of the overall system modelling and will thus enable design decisions and test considerations for software to be addressed within a system context.

In summary, hazard analysis may be used at an arbitrary level of application, both for hardware and software. For software testing, it may provide the basis for simulation both at an environment level – in terms of failure mechanisms and hazard scenarios – as well as at the detailed software specification level – where the requirements for justification of safety-related components may be identified and modelled.

2.4 The System Safety Case

In general, UK legislation requires the operators of hazardous installations to submit a safety case which describes how the design will provide for inherent safety, the systems to ensure safe operation and the means of verifying the design and construction quality. The safety case would also address installation activities and the limits of safe operation. The safety case is typically presented as the record of all the safety activities carried out throughout the project lifecycle. It is often a multi-volume document and presents detailed arguments for the safety of the system. As such the primary purpose of the safety case is to demonstrate adequate safety in design and operation. The safety case is a tool to manage safety, not just a paper exercise.

The requirements for safety case documentation have been defined in DEF STANs 00-55 and 00-56 [13], [14] and include the following:

- descriptions of the safety-related system
- evidence of competence of personnel involved in any safety activity
- specification of safety requirements
- results of hazard and risk analysis
- details of risk reduction techniques employed
- results of design analysis showing that all the required safety targets are met
- results of all verification and validation activities
- records of safety reviews
- records of any incidents which occur through the life of the system
- records of all changes to the system and justification of its continued safety.

In a wider sense, the safety case attributes should also include:

- reference to organisational procedures and guidelines
- reference to review procedures
- reference to a proposed lifecycle time scale and development, construction, operations, maintenance and decommissioning requirements
- reference to the organisation's project management arrangements.

It is important that the safety case is maintained throughout system development and also during its operation. It must show how system safety is maintained as functional requirements, design and possibly even tolerable risk and therefore safety targets change. The safety implications of any modifications must be completely justified in the safety case.

2.4.1 Safety Case Documentation

In general a safety case will include the following:

- a description of the plant or system
- hazard identification of normal operations and fault conditions, and hazard analysis to predict the potential consequences and their likelihood
- descriptions of the safeguards (engineered systems and administrative controls) that prevent, control and mitigate the hazards
- an assessment of the adequacy of the safeguards and consideration of the level of residual risk, to demonstrate that the risk is ALARP (As Low As Reasonably Practicable).

It should be noted that the proposed contents are generic in the sense that they provide a broad description of the information which may be developed within a safety case. The decision regarding which particular aspects of these contents are actually developed and the depth and form of their development will be made in accordance with industry and project specific criteria. These will be dependent upon wider risk criteria, which will take into account perceptions of tolerable risk.

The systematic development of the safety case will provide the basis from which the safety aspects of a system are clearly defined and the requirements for testing may be objectively established.

The Safety Case: Generic Contents

The proposed contents of a generic safety case is given below. This is intended as guidance on the issues which may need to be addressed in preparing a safety case.

Scope of The Safety Case
- Extent of consent sought
- Location of plant and physical boundaries
- Interaction and interfaces with other plants

Safety Standards and Criteria
- Safety principles and design standards
- Risk criteria

Description of System
- Process description
- Justification of process selection
- Plant description
- Modes of operation
- Supplies and services
- Control and Instrumentation philosophy
- Role of the operator
- Design, construction, commissioning, operation and maintenance

- Quality assurance
- Review of previous assessment documents
- Critical reviews of plant state and the operating experience.
- Outstanding issues and remedial work

The Safety Management System
- Company safety policy and objectives
- Organisation, staffing and responsibilities
- Qualifications and experience requirements / training
- Instructions and documentation
- Supervision and control
- Emergency systems
- Monitoring and auditing procedures
- Management of change and modification

Safety Assessment
- Safety assessment methodology
- Hazard identification
- Hazard analysis
- Sensitivity analysis
- Summary of safeguards and hazard control

Justification of Adequacy of Safety
- Normal operations
- Fault conditions
- Conclusions

Supporting Information This section may contain supporting information such as HAZOP reports, references, audit reports, etc.

Data Relevant to Future Review This section may contain information relevant to the future formal revision of the safety case such as modification to the plant and operations, new risk assessment methodology and changes to the safety standards and the risk criteria.

2.4.2 Software Considerations within the Safety Case

Within the UK there is no single mechanism for regulating the development and operation of safety-related systems which employ software. Regulatory mechanisms vary widely from industry to industry with, for example, voluntary regulation in medical engineering, regulation through insurance agencies and the HSE in the process and transport industries, and system certification through independent agencies in the avionics and nuclear industries.

Despite the lack of standard regulation across industry, a standard practice is evolving with regulatory authorities and enforcement agencies requiring that a safety case is developed which covers software and which shows both how the required level of software safety integrity has been determined for a given system and how this level has been achieved in practice.

These safety cases are assessed by the regulators/agencies and only if the arguments and evidence presented adequately support the level of safety integrity being claimed are the developed systems permitted to proceed with development and finally commence operation. As a minimum regulators/agencies normally require that a recognised standard for safety-critical software systems (such as DEF STAN 00-55) is followed throughout system development.

Currently within industry there is no generally accepted definition of what the software aspect of a safety case should include nor the development activities required to support it. There are however some existing and emerging standards, for example DEF STANs 00-55 and 00-56 [13], [14] and the IEC 1508 draft standard [6], which specify safety management requirements for software based systems and identify safety records that are required to be maintained to support justification of the system. These safety records effectively represent the basis of the safety case documentation.

To appreciate the essential elements for the consideration of software within a safety case, we must consider the activities over and above the conventional software development activities which are necessary to assist the objective of achieving a safe system. These include:

- hazard and risk analysis
- specification of safety integrity targets
- identification of risk reduction facilities
- safety functions definition
- safety compliance assessment
- safety verification
- maintenance of safety records.

Hazard and risk analysis needs to be carried out to identify potential hazards in the system, both internal and external, and the associated event sequences. The risk associated with each hazard has then to be estimated and safety integrity targets determined for the system. These targets are based upon the tolerable level of risk for each hazard. As they may well be different from the assessed levels of risk it will often be necessary to undertake a risk reduction exercise, by defining special risk reduction facilities to ensure that the tolerable risk levels are achieved.

Throughout system design and development of the safety-related system a continuous assessment is required to ensure compliance with the safety requirements of the system. A programme of safety validation should be implemented to provide confidence that the claimed integrity level is being achieved in practice as well as in principle. This may include independent assessment and validation.

2.4.3 Software Safety Testing within the Safety Case

Software safety integrity must be considered from a system perspective and, as with system safety integrity, performance standards should be set with reference to the risk reduction duty to be performed. The justification of safety claims will be based upon evidence which will typically include:

- credible arguments about the adequacy of the requirements specification, in terms of the hazards to be addressed
- credible arguments about the ability of the design and implementation to satisfy the safety requirements
- credible arguments about the adequacy of the development process in terms of its ability to avoid the introduction of faults and to detect and remove faults
- empirical evidence of the systems ability to satisfy the safety requirements (results of testing).

Safety integrity is thus concerned with both product and process issues and these concerns are extended from a system to component level, including software. Testing to support safety claims is necessary but not sufficient in itself as it will provide only one strand of the evidence required for safety justification.

The development of the requirements for testing the safety aspects of a system will be generated from this body of evidence contained within the safety case. The safety case fully describes the safety aspects of a system, including the requirements for and results of safety testing.

2.5 Lifecycle Issues

Whilst the stages by which a safety case will develop will vary in practice, the definition of formal milestones during project development will be required to support the interaction of safety activities with the development activities. Such milestones may also be a requirement of regulatory or certifying bodies. This section considers lifecycle issues for safety case development, as a basis for defining relevant activities and emergent products against the main project lifecycle.

Idealised lifecycle models present an abstract view which may not accurately reflect actual project practice. This is also the case with safety lifecycles, where the content and development of safety cases may be industry and possibly project specific. Any presentation of safety lifecycle phases must therefore be limited in that it will represent a particular instance of how the body of information in the safety case, together with associated activities and information, may be developed.

Safety may be considered as a distinct strand of the overall system design and development lifecycle. As such, it will involve its own activities and features, and interact with other strands of the overall lifecycle in a project specific manner.

2.5.1 Standards Approach to Safety Lifecycles

The management of safety requirements is an integral part of the development lifecycle. Project management should identify the processes involved in developing the safety-related system and choose a suitable lifecycle model. Standards such as IEC 1508 draft standard [6] provide detailed recommendations for processes that should be employed for developing safety-related systems of particular safety integrity level.

The lifecycle model should emphasise the importance of managing the functional safety requirements as an integral part of the development process. In the IEC 1508 draft standard [6] the overall product or system lifecycle incorporates a safety dimension which is referred to as the *overall safety lifecycle*. This is illustrated in Figure 2-2.

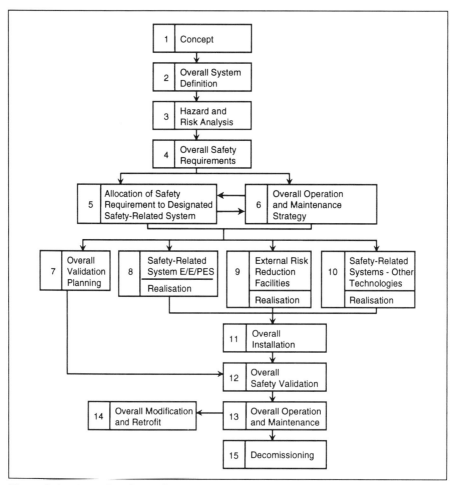

Figure 2-2 : Overall Safety Lifecycle

This figure is reproduced from [6]; © British Standards Institution.

Hazard Log and Safety Cases

DEF STAN 00-56 [14] provides guidance for maintenance of hazards data. It recommends that a *hazard log* should be maintained for safety-critical systems that will act as a principal reference document for demonstrating the safety characteristics of the system, and will provide traceability of the hazard management processes. The safety case document may contain detailed descriptions of processes and procedures involved in achieving safety, and the hazard log is primarily used for maintaining hazards and hazard related data throughout the development operation and maintenance lifecycle. DEF standard 00-56 [14] states that:

> *'The Hazard Log should act as the central control and reference document for demonstrating the safety characteristics of the system, and should provide traceability of the hazard management process. The Hazard Log should act as the focus of the logical argument, or Safety Case, for the deployment of the system into service.'*

As we are mainly interested in the development process of safety-related programmable electronic systems (PES) box 8 of Figure 2-2 has been expanded in Figure 2-3. It can be seen that the realisation of the PES indicated in box 8 has its own safety lifecycle. The IEC draft standard also considers the relationship between the software and safety lifecylces. This relationship is illustrated in Figure 2-3.

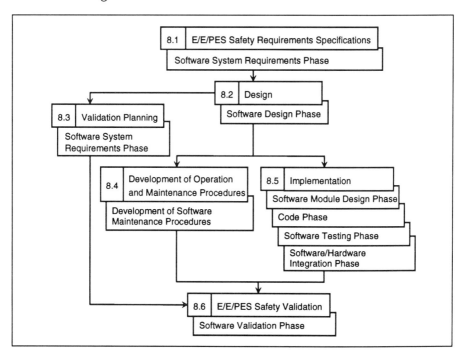

Figure 2-3: Relationship Between Software and Safety Lifecycles

This figure is reproduced from [6]; © British Standards Institution.

2.5.2 The Project Safety Case Lifecycle

Safety assessment should commence as early in the project as possible and develop with the project as detailed information becomes available. In this way, the identified hazards may be eliminated or minimised more readily and it is therefore possible to provide continuing assurance that the project will and does meet the design standards and the risk criteria.

In the nuclear industry for example, for plant that has more than a minor radiological safety significance, each stage of the project design, construction, commissioning, operation, and eventual decommissioning, should be covered by a safety case before commencement to the next stage.

An example of the relationship between the project life cycle and safety case development is illustrated in Figure 2-4, which is based upon nuclear industry practice.

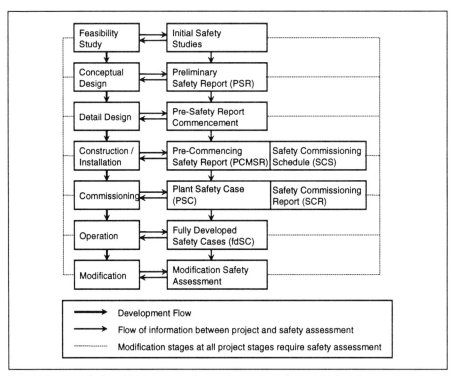

Figure 2-4: *Relationship Between the Project Lifecycle and Safety Case Development*

The content of a safety case needs to be such that all relevant safety issues are addressed. The level and detail of the assessment are dependent on the severity of the hazards associated with the plant. This categorisation of the plant, the systems and the components will also determine the standards of integrity to which those items should be designed, constructed and operated, and the appropriate level of audit and quality assurance (QA) requirements.

2.5.3 The Dual Lifecycle Model and Safety

The safety lifecycle plays an integral role throughout the system lifecycle. If simulation is used to validate a safety-related system then it is useful to consider a modified V lifecycle model known as the dual lifecycle. This is based on two conventional V lifecycle models, one for the system, and one for the simulator. In this section the processes that should be included in this lifecycle model to assist the integration between software development and the safety case are identified. These processes are generic, so that they are applicable in a wide range of industrial sectors.

Figure 2-5 shows the relationship between various phases of the dual lifecycle and the safety lifecycle of PES based systems shown in Figure 2-3.

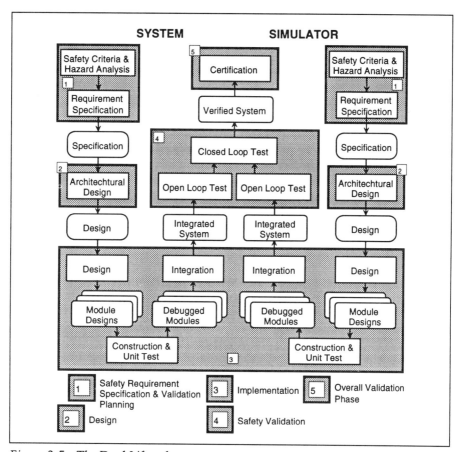

Figure 2-5: The Dual Lifecycle

The activities to be carried out during each phase of the development lifecycle are likely to differ between the system and the simulator and will be dependent upon the required safety integrity. In the case of high integrity systems in particular, if environment simulation is to be used for validation of the system then a separate lifecycle for the environment simulator is recommended.

The timescale for the simulator development is usually much shorter than that for the system under test. If the development activities shown in Figure 2-5 are mapped on to a time line as a Gantt chart, it becomes clear how constrained the simulator lifecycle is.

Figure 2-6 shows that the final requirements for the simulator are not known until the requirements specification for the system has been accepted, and the simulator development needs to be complete and the simulator validated before it is used in the closed loop test of the system.

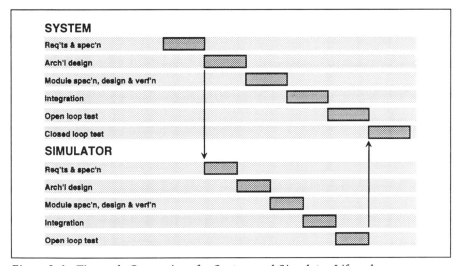

Figure 2-6 : Timescale Comparison for System and Simulator Lifecycles

Safety Requirements Specification and Validation Planning

These phases are shown in the upper part of Figure 2-4 They encompass the safety criteria selection and hazard analysis, and the requirement specification phases of the product and the environment simulator lifecycle models. The hazard and risk analysis should be performed at several levels of the development lifecycle starting from the Overall System Definition phase in the safety lifecycle. All hazards and event sequences leading to the hazards must be identified; the risk associated with each hazard must be specified using either qualitative or quantitative methods.

It is generally recognised that safety assessment should be started during the earlier phases of the development lifecycle. In these phases the documents to be produced may include:

The Safety Management Plan

The System Validation Specification This document will contain the Electrical / Electronic / Programmable Electronic Systems (E/E/PES) safety requirement specification from which we may justify that each safety function meets the safety integrity requirements. It will also provide details of the following:

- specific references in the E/E/PES Functional Requirement Specification
- the test environment (If environment simulation is used for validation then it will reference the simulator requirements and design documents. It will also reference the simulator verification and validation specification documents.)
- the procedures to be applied to validate correct operation of each safety function, and pass/fail criteria for accomplishing the tests.

The initial safety case documentation some of this information will be referenced from the System Validation Specification document.

Most of these documents are interrelated and the information may be referenced from one document to another. Initially these documents will be in outline form and as the project progresses into the implementation phases the contents of the documents will be detailed.

A formal requirements specification for the environment simulator and agreement of this specification is only recommended when the simulation result forms a major part of the safety cases or where the simulator is itself a deliverable item (e.g. the simulator may be used for training operators).

If an environment simulator is to be used to demonstrate the safety of the system then the environment simulator must be developed using development procedures commensurate with those used for the main system development. However some of the integrity requirements e.g. reliability/availability could be relaxed if necessary. The requirements for the environment simulator should be defined from the system requirements and the System Validation Specification. The safety case documentation should contain sufficient evidence to justify safety claims based upon simulation results. The requirements analysis of an environment simulator will include identification of the event sequences/responses to be simulated and the expected responses from the system.

Design

The overall design phase will encompass the design phases of the product and the environment simulator. The design of the E/E/PES must meet the functional and the safety integrity requirements of the requirements specification. During the design process all hardware and software interactions must be identified and documented. The safety-related items must be identified and safety integrity level must be assigned. The system architecture must be chosen carefully, if safety functions and non-safety functions are to be implemented in the same design then methods of achieving the degree of independence required during design must be specified in the design documentation. The IEC 1508 draft standard [6] recommends that if the design is based on a decomposition into subsystems then each subsystem should have design and test specification documents. The test specification documentation may include the following information:

- the types of test to be performed – this is related to the assessments performed by the assessors; who may wish to perform tests at several levels. The test objectives should be clearly identified and agreed upon by the developers and the assessors
- the test environment, tools, configuration and programs – the test environment should be identified and the tools used should be validated and verified according to agreed standards. If environment simulation is used to demonstrate the safety or functionality then it should reference the procedures involved in developing the simulator
- the test criteria on which the successful completion of the test will be judged.

Implementation

This phase encompasses the module design and construction, integration and open loop test phases of the product and simulator lifecycles. The procedures involved in developing the simulator and the products have been discussed in Leveson and Harvey [21]. Prior to this phase, the requirements specification of the product and simulator have gone through successive decomposition; and the derived requirements will be implemented in the software and the hardware modules. The processes involved in designing the system may be described in the safety case documentation under the section heading of *The Description of Plant and Process*. Each system level safety requirement can give rise to a large number of software requirements; in this phase the use of requirement traceability tools is highly recommended.

The E/E/PES will be implemented according to the design specification that is documented in the design documentation and will be tested according to the test documentation. If the subsystem implementation is different from that specified in the design documentation but meets the objectives then the differences must be justified and supported with test results. The draft IEC 1508 draft standard recommends that a test report should be produced that will state the integration and test results. The test report should include the following:

- the version of the test specification used
- the version of the software and hardware tested
- the test environment, test data, tools and equipment
- the result of each test
- the discrepancies between expected and actual results.

Safety Validation

This phase consists of the open loop test phases for the environment simulator and the product and the closed loop test phase. Environment simulators usually have a lower integrity level than the system. However, in some applications the failure of the simulator to detect fault in the product during testing may endanger the safety of the plant and the operators. In such cases the simulator must be validated using rigorous procedures prior to its use. Not all of the functions of the simulator may be safety-related; the safety validation report for the simulator should provide details of the validation procedures including the test environment and test results for each safety function. For simulators of higher integrity levels (levels 3 and 4), a separate simulator may be required to test the environment simulator. The components (interface modules) of the second simulator may be parts of the system.

The system validation process must be carried out according to the system validation specification as discussed in the earlier sections. A safety validation report should be produced which should provide the following details for each safety function:

- the safety function under test
- tools and equipment used
- the results of each test
- discrepancies between expected and actual results.

The safety validation report should be subjected to analysis and approval. During this phase the regulatory authorities may demand evidence of validation and verification, and the safety case will be used to demonstrate the safety of the overall system. The role played by the regulatory authorities will be dependent on the particular industrial sector and the integrity level of the product.

2.6 Guidance

> **Guidance on Testing and the Safety Case**
>
> - Ensure that the safety and system development lifecycles are documented and interlinked.
> - The environment lifecycle and the testing activities which form part of the overall System Development Lifecycle need to be interlinked.
> - The hazard analysis should encompass the software items within the system.
> - Determine the required safety integrity level for the software and document the hazards in which the software may be implicated. (e.g. in the hazard log.)
> - The safety integrity level of the environment software and its level relative to the safety integrity level for the system under development needs to be determined.
> - Maintain a well documented safety case from the outset of the project through to, and during, operational use.
> - Perform software safety assessments at defined stages, e.g. at the requirements, design and commissioning phases of the project.
> - Ensure that the risk reduction measures and safety function implementation are adequately monitored.

Table 2-1 : Summary of Hazard Analysis Techniques

Technique	Prime Use	Main Outputs	Use in Software System Lifecycle
1. HAZOP (Hazard and Operability Study)	Identification of hazards, areas of uncertainty.	Actions which should define a 'problem' more rigorously or provide information that allows production of a solution to a problem. Increase the level of awareness of the project, and how its objectives will be reached. Allows engineers/ scientists of different disciplines to learn together and helps forms a team; the output of which may be greater than the sum of the individual parts.	As an identifier to define more fully the extent of a problem or area of uncertainty. Most obvious use would be at the beginning of a project but periodic use at other stages in the lifecycle brings benefits.
2. FMECA (Failure Modes, Effects and Criticality Analysis)	Identification of hazards at component level. Generally a FMECA study is commissioned by the HAZOP leader.	Set of failure mode descriptions at component level. Relevant conditions are included e.g. Hazardous atmosphere, compatibility of material. A ranking of hazards and failures to allow targeting of resources.	As for HAZOP above.

Table 2-1 : Summary of Hazard Analysis Techniques (Continued)

Technique	Prime Use	Main Outputs	Use in Software System Lifecycle
3. ETA (Event Tree Analysis)	A 'bottom up' approach to identify the interrelationships between basic events and unwanted consequences.	Sets of 'logic trees' which show the relationship between an error and the undesired outcome. A list of identified errors and consequences. When reviewed or input to 'causal net analysis', sensitive areas of the software may be highlighted.	At an intermediate stage in the development life cycle to show relationship between types of errors and likely undesirable outcomes.
4. FTA (Fault Tree Analysis)	The opposite of an Event Tree. A 'top down' approach that allows an undesirable outcome to be analysed down to its basic causes.	Sets of 'logic trees' which are mirror images of ETA trees. A fault can be identified and a logic route to its most likely cause can be developed. Probability values can be assigned to each basic event and the probability of the fault occurring can then be quantified.	At an intermediate stage in the development life cycles to quantify the effect of interrelated failures and to track the cause – effect relationship between software errors and their consequences.

Table 2-1 : Summary of Hazard Analysis Techniques (Continued)

Technique	Prime Use	Main Outputs	Use in Software System Lifecycle
5. CCA (Cause Consequence Analysis)	A combination of ETA and FTA which has a starting point somewhere in the middle of a complex set of relationships.	A graphical output which may have more than one 'top event' if desired. It enables a more comprehensive view of cause and effect relationships than ETA and FTA individually.	Towards the latter stages in the development life cycle. Could be used as a check procedure after ETA or FTA. Use as management overview of the software/hardware interaction.
6. HEA (Human Error Analysis)	Analysis of human error on system operation.	Details of human error for system operation functions. Quantitative assessment of human failure modes so that sensitive functions are highlighted.	Toward the latter stages of the life cycle. Historically, HEA is a specialised technique which analyses the human tasks associated with system operation. Quantitative error factors could be used to assess the sensitivity of software interactions and consequences of human error.

Chapter 3
Designing for Testability

Safety-related systems should be designed for testability. This chapter focuses on real time control systems, and considers some typical design and implementation approaches. The emphasis is on the merits of these approaches for testability. The introduction to this chapter describes some typical safety-related applications. The merits of hardware and software interfaces between system functions is discussed, followed by a discussion of implementation options and software issues that affect testability.

3.1 Introduction

It is an impossible task to consider the complete range of applications and implementation options for safety-related systems. Testability should be carefully considered when the system is being designed. Some examples of safety-related applications are discussed below and implementation options and software issues that affect testability follow in the remainder of the chapter.

Many safety-related control systems incorporate modulating closed loop control of plant variables. There is a vast range of techniques for the design of the dynamic behaviour of such control loops. Only the implementation aspects of control loops will be considered here.

3.1.1 Single Input Single Output Control (SISO)

For Single Input Single Output Control (SISO) loops there are many special purpose instruments that perform closed loop control and which can be linked to a supervisory system via a communication channel. These instruments can operate in stand-alone mode and have auto and/or manual switch over. This allows the instrument to be commissioned independently of the remainder of the control system. It is common for these instruments to be controlled by the supervisory system via the communication channel. Typical functions include the ability to switch the loop from auto to manual, to change the controlled variable through its range and to monitor and tune the control algorithm constants. The testing of these functions usually requires the use of serial communications software.

In Programmable Logic Controllers (PLCs), controller blocks are supplied as separate software modules that can be enabled by the ladder logic. In this case, the testing within the PES is one of a module initiation test.

Regardless of its form of implementation, it is preferable to test the performance of the controller within its control loop by measurement of its inputs and outputs using analogue recording equipment. The demanded value or set point may not be available for direct measurement. This would be required if the control loop has a step response performance specification. Frequency response requirements usually involve some disturbance rejection requirement and so an appropriate signal representing the disturbance must be injected and the system response must demonstrate the specified attenuation. Signal analysis equipment/software will be required. In either case it will be necessary to provide some means of simulating the plant response to control signals in order to close the loop. The accuracy of the simulation model will determine the validity of such tests.

Control loop performance is dependent on both the PES and the plant; it is not the sole responsibility of the PES to deliver the required performance, the plant must be designed so that the required performance is achievable.

3.1.2 Multiple Input Multiple Output Control (MIMO)

Multiple Input Multiple Output Control (MIMO) control algorithms have been proposed for systems that have closely coupled loop dynamics and where there is a requirement for a reduction in interaction between controlled variables. Many examples of simulated application of such controllers in various application areas may be found in the literature. However, the integrity of such control structures is difficult to guarantee by theory or analysis. For example, if one loop is opened, due to actuator failure for example, then stability of the system is not necessarily retained, even though the plant itself may be open loop stable. In order to check that a MIMO controller has integrity it has to be tested for the opening of all combinations and permutations of the loops, this can obviously lead to a large number of test cases. Furthermore, a realistic representation of the plant must be available in order to make such tests possible.

3.1.3 Control Loop Failure

Control loops usually have associated with them alarm and trip levels on the measured variables and in some cases on the control signal. In the event of the signal passing through one of these levels specified actions are required to take place. For alarm levels this may include logging of events, audible alarms, altering displays. For trip levels, actions may include opening of the control loop and maintaining the controlled variable at its current level or sending it to an upper or lower limit as dictated by the plant behaviour. There may be some requirement on the transient behaviour of the control signals following a trip dependent on the plant dynamics. The equipment required to monitor the PES response during a trip may range from storage oscilloscopes to chart recorders. Other events that may be signalled to the control software by digital inputs may elicit similar responses.

In some systems there may be a requirement for 'graceful degradation' of control. This can be interpreted as a sequential shutdown of control on areas of the plant, possibly on 'healthy' loops. There may also be a requirement for auto restart on clearance of all fault conditions, i.e. the initiating fault and all subsequent faults that would have tripped the system, had it not already been tripped. Both these requirements introduce more complexity in pre-conditions and post-test conditions.

3.2 Architectural Considerations

One of the major decisions available to a system designer is the system architecture, i.e. the form in which the hardware is arranged to support and supply the intended functions.

3.2.1 Segregation of Function

Given certain input conditions the PES must react in a specified manner depending on its state. The software that controls the transition of the PES from one state to another is termed the control logic. To test state transitions the PES must be driven to its pre-condition state, the event causing the transition must then be applied and the post-condition must be confirmed. Using the structured design approach the control logic is represented as finite state machines, the output of which control the initiation of processes which handle a variety of data.

A range of CASE tools is available for specifying control logic and animating the resulting specifications. These tools support the prototyping approach to system development and provide a halfway house between formal methods and logic gate diagrams. This segregation of Boolean operations and mixed data operations lends weight to the segregation of system functionality by use of separate hardware items. For example a common PES architecture is one in which the control (protection) logic is implemented in a PLC while the modulating control may be implemented within a process controller. The PLC determines when and if a process controller should be active. Thus testing of alarm and trip conditions and logic is focused on an identifiable section of the PES.

An example of a typical PES architecture is shown in Figure 3-1. The use of the PLC-Process Controller arrangement is attractive as it allows the testing of the control logic and modulating control functions in isolation. The function and hardware become closely linked and thus the hardware operation can be optimised for its intended function. Units can be replaced if abnormal behaviour is observed during testing to ascertain whether the behaviour is a unit or system problem. This architecture provides hardware interfaces

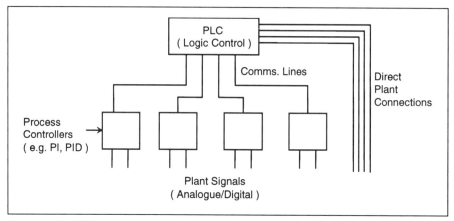

Figure 3-1 : PES: Architecture

between functions that can be used for test purposes during integration testing if so desired. The recommendations from the standards and guidelines on architecture are discussed in Section 1.6.3.

3.2.2 Multiple Channel Architecture

It is common in high integrity applications for a multiple channel PES architecture to be adopted. If there are differences between software in each channel, each channel must be tested individually unless good arguments can be put forward to the contrary. The arguments that may be put forward are similar to those that would be put forward for limited retest following a change to a single channel system. The voting logic associated with multiple channel PESs is normally implemented using hardware.

It is very likely that self test functions are included in each channel of a multiple channel protection system. It is necessary to be able to filter out the effects of self tests when performing external tests. This requires that dedicated indications of on-going self tests are included in the system design.

3.3 PES Interface Considerations

The merits of a hardware interface as against a software interface between system functions should be considered. A number of positive and negative points for each type are listed in Table 3-1.

It can be seen that the positive and negative factors associated with each class nearly balance, therefore it cannot be said that one class of interface should be used in preference to another. The advice must be to bear the points above in mind when designing the system and take account of those that are most applicable to the particular system. There may also be other fac-

Table 3-1 : Software and Hardware Interface Test Merits

Software Interface	Score	Hardware Interface	Score
Application specific	−	General	+
Monitor will interfere with system operation	−	Monitor need not interfere with system operation	+
Can handle large number of variables at low cost	+	Hardware costs may restrict variables monitored	−
Timer resolution limited by application system	−	High resolution timing equipment available	+
System failure halts monitoring	−	Monitoring equipment independent of system	+
Monitor can be controlled by system under test	+	Extra control lines expensive and intrusive	−
High speed data transfer	+	Data transfer may be limited by hardware	−

tors that will influence the selection of a hardware or software interface, such as safety.

From the point of view of testing high integrity systems the non-intrusive nature of hardware monitors is attractive. The use of a single hardware/software platform does not provide the usual accessibility to interfaces, indeed the hardware interfaces may not exist, all the processing being performed on a single processor. In such situations individual modules may be tested in isolation and possibly grouped in subsystems before the complete system is transferred onto the final target hardware. Figure 3-2 shows some of the principal hardware interfaces within a typical system.

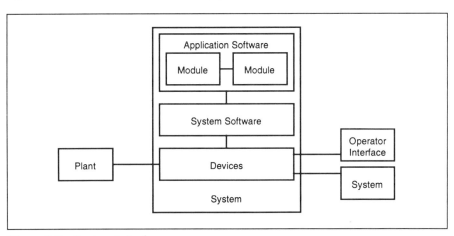

Figure 3-2 : System Interfaces

3.4 Implementation Options and Testing Attributes

The choice of implementation option and its effect on dynamic testing is discussed in the following paragraphs.

It is worthwhile at this point to review the objective of the test process. For functional testing the aim is to demonstrate compliance with a specification. In order to do this the specification will have been interpreted to produce test cases. Test cases will have pre-conditions and post-conditions. Thus we must establish and observe the pre-conditions and observe the post-conditions. We must know when the pre-conditions are met so that we can start execution of the process and know when the post-conditions should have been met so that we know when to record the test results. Establishing pre-conditions during testing will depend on the controllability of the system under test, which will in turn depend on interfaces provided and on the logic implemented within the system.

One approach, suggested in DEF STAN 00-55 [13], to testing a system beyond its positive requirements is to use random inputs with the system in operation. Both the timing and levels of signals may be random. With such testing it is therefore possible to exercise the system in an unrealistic manner, in that combinations of signals being randomly distributed do not fall within a 'typical' operational envelope. This may be seen as an advantage in that the system is being tested outside its normal operational envelope and that the system's sensitivity to corruption of signals may be investigated: an alternative view is that the distribution of the plant signals may be a physical impossibility and that the test has no useful interpretation.

For the purposes of this discussion, attributes of implementation options are considered to be the same across all application domains.

Real time structured analysis conventions for control and data flow models are used to describe the behaviour of the system. The aim of generating these simple models is to identify how, in broad terms, the implementation method and system behaviour affect the ability to perform tests.

3.4.1 Dynamic Testing Attributes

Figure 3-3 illustrates the typical relationship between a monitor process and the system under test.

The monitor process must be able to observe the input and output buffers and the system memory. It must also be able to export the observed data to some external equipment, labelled as analysis equipment. The testability of the system is very much determined by the performance of the monitor and its interface to the external analysis equipment. The speed at which the monitor can execute determines the resolution of observations on the system under test. There are two modes of operation for the monitor process, it may operate as a sampler or it may run as an event trapper.

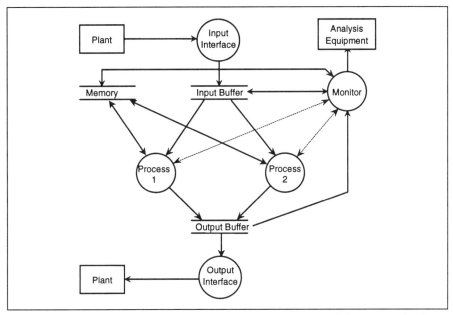

Figure 3-3: Typical Relationship Between Monitor Process and System Under Test

As a sampler, the monitor must scan input and output buffers and memory and transmit data to the analysis equipment at predetermined intervals. Implementation of such a monitor embedded within the target system is undesirable as it will affect the performance and resource utilisation, possibly well in excess of that required by the application. An alternative would be to design the system so that the data stores indicated are observable by an external monitor. Dual ported memory would be one way of allowing access to data at a suitably rapid rate.

As an event trapper, the monitor must observe the input and output buffers and memories and registers very quickly in relation to execution of code. It compares the state of the data stores with a state of interest, if it detects the state it transmits the data to the analysis equipment. In large systems it may not be possible for the monitor to scan and mask the system data stores before they change. In order to reduce the amount of data to be scanned by the monitor, it may be possible to select only a subset of data stores that are regarded as critical to the observation of an event.

Clearly there is some need for synchronisation between the monitor process and the process or processes under test to allow unambiguous results to be recorded. There must also be careful consideration of signal, process and monitor dynamics, so that realistic targets can be set. Ideally the monitor process would be able to control execution of the processes under test. Control of execution by the monitor is desirable for a number of reasons: for example it would aid integration testing as processes could be loaded and 'switched-in' as each process is tested in conjunction with others. It may be desirable to place breakpoints in the process under test to allow the monitor to observe intermediate values. Or even to advance or retard one process with respect to others (in an asynchronously operating system) in order to investigate the effect.

The following attributes are defined to formalise concerns relating to execution and monitoring of dynamic tests and form the basis of the comparative discussion of various PES implementation options.

Controllability There is a need to consider whether the type of implementation aids or hinders the ability of the tester to inject signals that drive the system to the desired state.

Observability It is necessary to consider whether the type of implementation makes it difficult for the tester to record the signals required to provide evidence that the system is in a desired state.

Traceability The way in which the system is implemented may make it difficult to record the signals necessary to demonstrate that a particular path has been executed. It is assumed that adequate 'instrumentation' has been included in the process logic to ensure that unique evidence is available in the data areas available to the monitor.

Predictability We need to consider whether the system's mode of operation provides for a means indicating when execution of logic is complete, so that we know when to observe the outputs.

Repeatability The system may be sensitive to slight variations in timing. However, in this discussion the variation of functional behaviour of the system with slight variations in timing is not being considered, rather the effect of timing sensitivity to the testing process.

Functional Testability The ability to inject and record the signals associated with the input to output transformation of a functional 'block' needs to be considered.

Accessibility The ease with which signals from a test environment may be introduced to the items under test is important.

Four main implementation options for real time systems are now evaluated using the testability attributes discussed above.

3.4.2 Test Attributes of Various Implementation Options

Type A: Synchronous Operation Single-Tasking

Description of Behaviour Figure 3-4 represents the operation of a synchronous system. Each of the processes indicated on the diagram operate in sequence. The input interface is signalled to start a scan of inputs, placing the captured data in an input buffer. Once all inputs have been scanned, the control process is signalled. The input interface then waits for a signal to restart. The control process then signals the application process to commence. The application process operates on the data in the input buffer and any required data held in memory. As each step in the process completes, data is written to the output buffer. Once the application process reaches the terminating statement, it signals to the control process that it has completed. It then waits for a signal to restart. The control process then signals to the output interface to commence operation. The output interface takes data in the output buffer and converts this data into a form that can drive the plant. Once the output interface has processed all data in the output buffer it signals to the control process that it has completed. The control process then usually waits for a signal to restart. This signal is usually triggered after an elapsed time. This type of operation is typical of programmable logic controllers and digital process controllers.

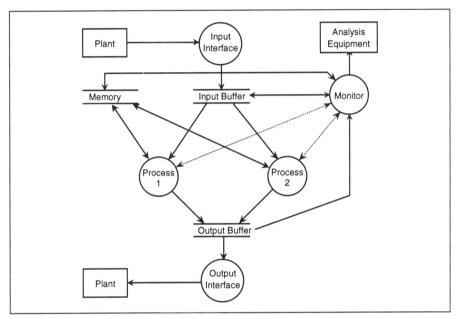

Figure 3-4: Synchronous Operation, Single-Tasking

Discussion of Test Attributes

Controllability Given that each system must accept input signals and that in driving the system to a desired state there is strict control of these signals, this mode of operation is not perceived to offer any particular advantages over type B, C or D systems, described later.

Observability Outputs change in a short well-defined time span, which make it easier to record outputs as they will be steady for a known period of time, the duration of the process execution time. The observability of the process logic is not considered here as it would be a coding and/or design issue particular to each type of system.

Traceability Recording inputs and outputs and memory contents at the start and end of the process execution will allow determination of the execution path (assuming adequate 'instrumentation' of the process logic).

Predictability The process under this mode of operation will execute at regular intervals. This will help in predicting when to record the contents of buffers and memory.

Repeatability The sampling and latching of the input interface at predetermined intervals desensitises the process logic with respect to timing of input signals. Provided they are applied at any time before the scan of inputs, the process logic will operate in the same way each time.

Functional Testability It is necessary to record the inputs as seen by the process logic to ascertain that the system has produced the correct transformation of inputs to outputs. As the inputs are all latched at the start of execution, the monitor is free to record them at any time before the next scan in the knowledge that the signals recorded are those employed by the process logic at all stages of execution. This means that it is not necessary to time stamp the inputs. This is seen as an advantage over type B and C systems.

Accessibility Because the system is single-tasking, it is not possible to run a separate task within the system that can manipulate the system memory. This makes the system less accessible than type C and D systems. The system could be run in debug mode whereby each instruction is executed and control returned to the debugger. Within the debugger one generally has the ability to manipulate memory. If the scan time of the application program is well within the capabilities of the hardware platform, it may be possible to embed within the application code calls to simulation subroutines that can manipulate system memory, these simulation subroutines would form part of the application software and would have to be disabled in the delivered system. In order to maintain timing of the 'as tested' system, delays would have to be introduced to replace the execution of the simulation procedures.

Rather than interleave the simulation with the application code, it would be possible to run the simulation procedures at the end of each scan, (again providing that there is sufficient spare processor time before the commencement of the next scan). The simulation procedure could manipulate memory using hard coded procedures. A more flexible approach would be for the embedded simulation routine to service a communications channel, interpreting incoming messages and manipulating memory accordingly. We should be very careful about adopting such an approach if the simulation is interleaved with the application process as it would in effect introduce asynchronous input to the system. This may be exactly what we wish to avoid, its avoidance being the primary reason for selecting such an operating system in the first place.

These methods may be unattractive in that they affect the application software, increasing its complexity and introducing the possibility of incorrect operation in the field (if the simulation procedures are erroneously switched in). To avoid such worries then the simulation must access the system via existing interfaces.

Type B: Asynchronous Inputs in a Single-Tasking System

Description of Behaviour   Figure 3-5 represents the operation of a system in which the contents of the input buffer may change during the execution of the application process. The mechanism that allows this to occur is not the concern of this discussion. In other respects the operation of the system is similar to that for the system represented in Figure 3-4. In order to make such a system operationally identical to that for Figure 3-4, the simple measure of making local copies of the input buffer within the application process would suffice. Memory constraints may prevent this measure from being taken.

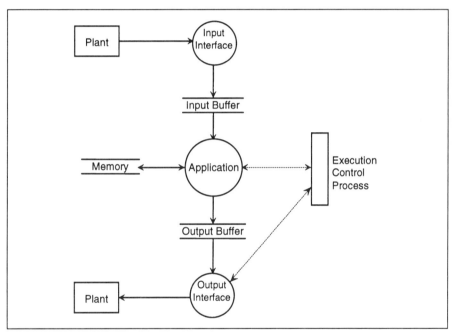

Figure 3-5: Asynchronous Inputs, Single-Tasking

Consider a process that uses a plant input for the basis of a number decisions. Using a system such as that shown in Figure 3-5 introduces the possibility of parts of the process using different values for that input if the input updates during the execution of the process. This may result in the generation of a set of inconsistent control signals that persist until all sections of the code have been executed using the same value of input variable. This could be a considerable time if the process has moved to a state where the 'inconsistent' branch is no longer executed. It should be clear that from the system security point of view such a method of operation is extremely undesirable.

Although it can never be guaranteed that the output changes are all made at the same instant, it is assumed that the execution of the output interface is very rapid when compared with plant dynamics. This would ensure that the time taken for the system outputs to update would be small thus reducing the time period during which there is a possibility of inconsistent control signals being applied. However, if outputs were allowed to change at any stage during the process execution, then the 'window of opportunity' for output mismatch would depend on the execution time of the process. As the branching within the process should not depend on outputs it can be argued that although asynchronous outputs may affect the performance of the plant, it will not affect consistent execution of the control process.

Given these doubts about the propriety of such a method of operation, the aspects of testing such a system will be discussed.

Discussion of Test Attributes

Controllability As for class A systems.

Observability As for class A systems.

Traceability In this case time stamped recordings of input data will be required so that synchronisation of input data with process execution, which must also be recorded, can be achieved during analysis of the test results. For example, if we are looking for evidence that a particular branch, which is dependent on a value of an input, was executed we would need to confirm that the input conditions for the branch were achieved; if they were not, due to the necessary signal being applied after the conditional statement being executed for example, it would be futile to search for the evidence.

Predictability As there is only one process there is little difficulty in ensuring regular execution and updates of output buffers.

Repeatability Due to the sensitivity of the process execution to the timing of signals it is considered that it is more difficult to obtain 'identical' conditions from test to test. If random input signals are applied to this type of system we are likely to get more variability in system behaviour and thus there will be a need for greater effort in interpretation of test results.

Functional Testability In order to ascertain that the system has produced the correct transformation it is necessary to know the data upon which the function operated. It is more difficult to collect the required data for this type of system than it is for a type A system (see Traceability above).

Accessibility As for a type A system.

Type C: Asynchronous Inputs with Multi-Tasking System

Description of Behaviour Figure 3-6 represents the operation of a multi-tasking system in which the input buffer may be updated during the execution of one or more processes. If the buffers and memory are well partitioned, then this system can be made to approximate to that of the system represented in Figure 3-5. It may be possible to load a single process into the target during module testing in which case, the system, although it has the capabilities represented in Figure 3-6, is configured in the manner shown in Figure 3.5. If the software is coded so that internal copies of the input buffer are created within the process then with the provisions mentioned above, the system can be made to approximate to that in Figure 3-4.

The processes could be executed in a round-robin sequence, each process running to completion before handing back to the scheduler. Each sequence could have the same priority, which would mean that the sequence of operation once defined would continually repeat. If each process could be assigned different priorities on-line, then the sequence of operation may alter and system operation would not be time deterministic. The principle of each process running to completion within a round-robin sequence would help in testing because each process could be regarded as part of an overall process.

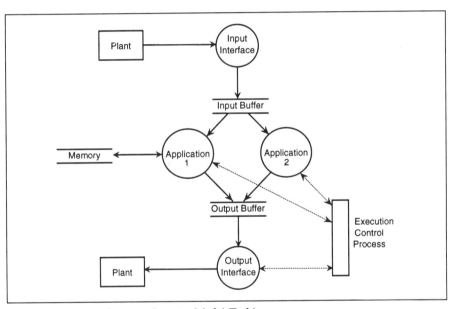

Figure 3-6: Asynchronous Inputs, Multi-Tasking

Consider a system in which the individual processes are allowed to free run without synchronisation. The processes may have different execution times and if there is no priority scheduling, the processes would cycle at different rates. This could create problems. Consider the situation in which process logic is fired by a transition in an input. 'Fast' processes would execute initially 'seeing' the transition and subsequently not 'seeing' the transition, while slower processes would still be executing 'seeing' the transition. This could lead to incorrect output patterns being applied until all processes had completed one cycle 'seeing' the transition. If the test specification required that the tester observes the outputs continuously these patterns may be observed. Alternatively, if the system outputs are sampled some time after completion of the slow processes a wholly consistent set of outputs would be observed, failing to observe that transient (possibly dangerous) conditions had occurred due to multiple execution of the faster processes. Clearly such issues should be catered for in the system design. From the point of view of testing, the aim of which should be to discover faults (Software Reliability, Myers [23]), the test specification should be couched in terms that would enable capture of any erroneous behaviour, thus an acceptable time frame for sampling test results must be specified.

One should appreciate that in any system there will be a short period during transitions when outputs may not be wholly consistent. As mentioned earlier, the duration of these inconsistencies should be short compared with associated plant dynamics, this includes any hardwired protection systems that monitor system outputs.

Rather than enforcing a round-robin sequence that may be too restrictive, synchronisation between interacting processes may be implemented. In multi-tasking systems synchronisation between processes can be achieved using semaphores or event flags. Semaphores are integers that can be initialised, decremented and incremented by processes. When the semaphore reaches predetermined values, processes monitoring the semaphore initiate actions. The monitoring processes need not be in the same set of the processes that set the semaphore's value. Thus if simulation processes have both read and write access to semaphores then they have the ability to control the execution of the application software.

Event flags are slightly less useful; by its nature only one process may set an event flag at any one time, but many processes may be initiated by the setting of the event flag. Thus simulation processes may monitor event flags to observe progress but, unless the event flag has been explicitly included for use by the simulation, its setting by the simulation could lead to spurious processing by the application software.

In Type C systems it is possible to allow the processes to run asynchronously. This is acceptable if the processes are well partitioned. Cost and compactness may be reasons for running independent processes in such an environment, but from the safety point of view, if the processes are independent then running a system that introduces the possibility of interaction means that extra testing must be undertaken to demonstrate that no interaction takes places where none was intended. One way of reducing this testing overhead is to decompose the PES, allocating specific system functions to separate processors. There are a number of advantages in taking this approach:

- the decomposition into safety-related and non-safety-related functions
- the opportunity to use diverse hardware/system software/application software
- the decomposition of the control problem into time-critical and non-time-critical operations.

The difficulties of testing a decomposed system include:
- the use of diverse hardware may introduce the need for diverse test environments
- the synchronisation of separate processor systems in establishing test pre-conditions.

Added difficulties of multi-tasking systems arise, even if synchronisation measures are implemented, due to asynchronous inputs affecting more than one process. If two or more processes use the same input then the instant of interrogation of the input buffer by each process and the time to cycle round each process becomes critical in assessing the execution of the code. If Process A uses input X at time t and Process B at time $t+\delta$, then the opportunity for a data output mismatch arises if input X changes in the interval $(t, t+\delta)$. If the cycle time of Process A is T_a and Process T_b then the duration of the mismatch would be, at least the greater of T_a or T_b. This should be set against the alternative, which is that the inputs are latched until all processes complete their cycle. In this case both processes would produce signals that could be up to T_a or T_b out of date. The system should be designed such that this interval should be small compared with plant dynamics.

Discussion of Test Attributes

Controllability As for type B systems with the addition of further processes to be controlled.

Observability The time span over which the output buffer is changed may be increased due to the different execution times of the processes. Provided that the output interface affects all output changes in one scan then this type of system is no more difficult to deal with than with type A or type B systems.

Traceability As for type B systems.

Predictability With the presence of multiple processes each with variable execution times due to scheduling activities, it is fair to say that type C systems are likely to be less predictable than type A or type B systems.

Repeatability As for type B systems with the added complication of many processes to monitor and control.

Functional Testability As for type B systems with the added complication of many processes, whose behaviour may have to be reproduced by the test environment.

Accessibility With a multi-tasking system comes opportunity to use a further process to provide a simulator. The simulation code does not have to be embedded within or linked with code under test. If global memory areas have been employed in the application software, then declaration of the global data within the simulation code gives full access to the application data. These features make multi-tasking systems more accessible than single-tasking systems (types A and B).

Type D: Requested I/O Operation

Description of Behaviour Figure 3-7 represents a class of system in which the input and output signals are under the direct control of individual processes. In such a system the process may issue a request to the I/O subsystem of the target operating system, the execution of the process will then wait for a signal from the I/O subsystem. For an input request it returns a signal to the process indicating that the data requested is ready to be read. For an output request, the signal back from the I/O subsystem may merely be an acknowledgement rather than a signal that the outputs change has been effected.

This type of operation allows the process to obtain the most up to date information on plant status before execution of statements dependent on that information. The input/output data operations execute synchronously with the process code, i.e. the process requests the information and waits for its return. On its return the process uses the data to complete its calculations. Using such a mechanism does not guarantee that the process executes synchronously with the plant. In a multi-tasking environment the time taken for each process to execute becomes indeterminate due to a number of factors, including the various execution paths available within each process, the current status of the task, a wait for an event or the priority of the task compared with others waiting for the CPU.

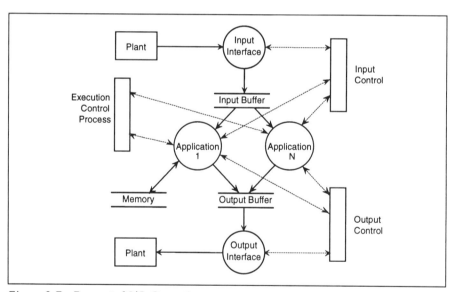

Figure 3-7 : Requested I/O Operation

Discussion of Test Attributes

Controllability As for type C systems.

Observability As outputs may be affected by requests from individual processes, rather than from a single controlling process, the outputs may change at indeterminate points in time. Knowing when to look at outputs is important and thus there is the added complication of a signal being required by the monitor to indicate that an individual process has completed and that outputs related to the process may be recorded. Other outputs unrelated to the completed process may be in a state of flux due to the execution of other processes and thus should not be recorded. There is obviously a need for a more sophisticated monitor that can select and record only relevant signals.

Traceability This type of system should be more traceable than type B or type C systems because the processes request input and wait for its arrival before continuing execution. Therefore we can observe the data areas at any time later sure in the knowledge that what we see now is the same data as the process logic used at some point earlier. Type B and type C systems allow data to be updated after processing possibly leading to confusion in interpreting results.

Predictability As for class C systems.

Repeatability Because the requests for input may be distributed in time the system's behaviour becomes more sensitive to timing of input signals than a type A system.

Functional Testability As outputs are under direct control of each process they can be expected to change at any time the process requests output services. This means that constant monitoring of outputs would be required if evidence of attainment of response times is required.

Accessibility As for class C systems.

3.4.3 Comparison Scores for the Implementation Options

On the basis of the discussions of earlier sections the attributes of each type of system have been assigned plus or minus marks relative to the other types.

It is accepted that such an analysis is subjective, but it provides a framework for making a case for use of one type of PES in preference to others.

It should be remembered that this analysis relates only to testability attributes and not to the appropriateness of a system type for a particular safety-related application. There are many programming techniques and implementation features which may ameliorate some of the less desirable aspects of system operation, some of these techniques and features have been discussed. However, the scoring of the mode of operation has been done with the assumption that these defensive techniques have not been implemented. On the basis of Table 3-2 Type A systems are deemed the most 'testable', this derives from their simplicity of operation and time deterministic behaviour.

Table 3-2 : Classification of PES Operation by Testability Attributes

Attribute	Class A	Class B	Class C	Class D
Controllability	+	+	+	+
Observability	+	+	+	−
Traceability	+	−	−	+
Predictability	+	+	−	−
Repeatability	+	−	−	−
Functional Testability	+	−	−	−
Accessibility	−	−	+	+

3.5 Software Features

Software should be designed keeping in mind the principle that testing requires controllability and observability. Clearly a software item must be able to be stimulated in some way to perform its assigned task otherwise it would not function. At the testing stage it is likely that many more variables are required to be monitored than will actually be required during operation. The problem of instrumenting code for test purposes without affecting its performance can be difficult.

The issues discussed here are: how easy is it to stimulate that function for test purposes and how can the behaviour of the software function be observed.

3.5.1 Hierarchical Structure Design

A system that is broken down into a number of discrete elements each of which performs a well-defined task has come to be regarded as the best way of managing the complexity usually associated with software based systems. A number of methods are available to ensure that the tasks in the system are well defined and consistent (IEC 1508 draft standard [6]).

Structured design allows modules to be written which perform specific tasks. These tasks can then be tested independently of the rest of the system so that it is clear that each task is being stimulated and the results can be easily observed. A well structured system also has the effect of reducing complexity and increasing the focus on the operations performed by the module.

The hierarchy of a software system should be documented. If a structured design approach has been employed then this information will be available in the structure diagrams as part of the supporting documentation. The structure diagrams enable the identification of software sub-systems that can be tested in isolation using bottom up integration and modules that have to be simulated as stubs using top down integration. The structure diagram can be used therefore as a source for planning of coding and testing activities.

The form of a software system structure chart will have an influence on the integration strategy adopted. In the extreme, with one module the 'big bang' integration is all that is possible. Some target systems, such as PLCs, effectively force a single module structure. Most programming languages now allow a modular structure and thus give the designers more flexibility.

The modular structure introduces the possibility of errors in the call statements between modules. This can be ameliorated by using a highly structured language that enforces clear variable naming conventions and type declarations.

Boehm [24] suggests that the effect on system test effort of the use of 'modern design methods' can be expressed by a factor on test effort ranging from 1.5 (no use of modern methods) to 0.65 (full use of modern methods), i.e. a reduction of over 50%.

3.5.2 Object Oriented Design and Programming

The benefits claimed for Object Oriented Design (OOD) and Object Oriented Programming (OOP) are of much the same general character as those offered by structured design and programming, but go much further in some directions, e.g. reuseability. As with any technology employed in safety-related applications, its use must be considered in the context of accepted safety principles. The most fundamental of these principles relate to simplicity and comprehensive validation.

Some aspects of OOD and OOP support safety principles, others may compromise them. It is a matter of judgement whether or not they are included in the combination of techniques for the development of a particular application. Some factors of OOD and OOP affecting simplicity and validation are identified in the following:

Reuse One considerable benefit claimed for OOD is reusability resulting from facilities for class hierarchies and inheritance. However, considerable resources, both in terms of design effort and cost, are needed to create reusable objects, with the main benefits accruing to subsequent projects. This implies that an object oriented based software system that is designed with reuse in mind may be more complex than required for the first system – something to be avoided in safety-related systems. Where objects are reused, validation costs should be lower although the problems of adopting software for use outside the operational envelope considered when it was originally tested need to be considered. The track record of object libraries, both in terms of maturity and scope of application, needs to be maintained.

Formalism Some OOP languages provide the ability to specify the formal properties that an object's operations must obey by writing assertions. Assertions may express pre-conditions, post-conditions or invariance conditions. These features are provided to help towards the goal of producing proofs of correctness and permit the inclusion of exception handling features.

Exception handling Although any run-time overhead associated with exception handling of assertions appears to be relatively low, such features seem to be most useful as a debugging aid during program development rather than as operational diagnostics. The occurrence of any run-time error in an operational system could have disastrous consequences – in some cases there would not be the time to recover. It is considered to be easier to write a program that is exception free, and prove it to be so rather than prove that the exception handlers take the appropriate action under all conditions.

Instantiation of objects The dynamic creation of objects hinders static analysis and review of code. Although it is possible to create static objects, the more pure the OO language used, the less possible it is to achieve. For any real time system, bounded memory and time requirements are the keys.

Language constraints Unless a pure OO language is used, it is easy to break the OO paradigm and compromise the inherent structuring and modularity. If this occurs it leads to increased complexity and ambiguity and potentially a loss of all benefits claimed for ease of modification.

Polymorphism The use of polymorphism may tend to confuse during code walkthroughs. The operation invoked is dependent upon the class of object receiving the message and the target object itself may be passed as a parameter.

Garbage collection Unless facilities are provided to control garbage collection it can result in non-deterministic real-time behaviour.

Order of Execution Within the OO design approach objects are conceived as being able to call upon the services of other objects asynchronously. It can be argued that this forces the designer to make the objects independent and therefore promotes the opportunity for reuse. It also tends to promote the use of multi-tasking to implement the independent objects. Multi-tasking is not favoured for safety-related applications.

3.5.3 Internal Complexity

The internal complexity of a module has a bearing on the amount of time required to perform tests and indeed the type of testing required. The less complex the code then the less complex the testing method requires to be. Generally, assessing complexity involves scanning source code for certain features such as number of lines of code, number of decision statements, number of loops, etc. These calculations can be performed either by hand or by automated tools for specific languages.

Complex modules can be decomposed into a number of simpler modules. The advantage being that it makes the function easier to test. The disadvantage is that it increases the coupling in the system and therefore makes the system as a whole more complex. By aiming for low coupling, the design may result in modules of greater complexity. A trade off between module coupling and complexity should be sought.

Internal coding standards may be amended to incorporate an acceptable limit on complexity as measured by a suitable metric, e.g. McCabe's.

3.5.4 Data Interchange

Perhaps a more measurable level of system modularity is the amount of data passed between modules. The less data that has to be passed between modules the more independent and focused they are. From the point of view of testing, the fewer parameters to be passed, the fewer signals that have to be simulated during module testing. Two ways in which data can be transferred between modules are via global data areas or as procedural parameters. There are merits to both methods.

Global Data Areas

Global data areas are areas of memory allocated to a set of fixed name variables and constants. They are declared in a separate code module and initialisation of these variables can be carried out.

A global data area is potentially accessible to all modules. This has obvious advantages for testing in that a test driver module could be built which has the ability to set and record the value of any variable in the global area. This means that the test module can monitor data at all levels within the software system.

The use of a global data area allows new functions to be added at any level without the need to modify modules other than the calling modules. This means that only the modules with changed function need be retested. Without the global data area all modules in the chain from source to point of use would have to be modified to pass the required data. All affected modules would have to be retested.

The potential disadvantage of using global data areas is that modules may have access to variables that they do not need, and so the chance of corruption of global data increases. This may cause problems interpreting static analysis results. One way of avoiding such problems is to make the structure of the global data area flat or by information hiding (IEC 1508 draft standard [6]). This ensures that declaration of external variables in modules can be limited to those required by the module. However the hidden data needs to be accessible to test harnesses. Special access procedures may have to be written for test purposes, this would clearly add to testing effort and may compromise the validity of the tests.

Language choice helps in limiting the visibility of 'global' data. For example in Ada, data variables may be declared in a package along with read/write procedures which deliver/receive data to/from procedures in other packages. The visibility of these read/write procedures may also be limited. Thus one has the advantage of keeping data in one place, yet at the same time one can limit access to the data to only those procedures that require it. This avoids 'tramp' data that is passed through a number of procedure calls without modification.

A disadvantage of using a global data area is that code reuse may be difficult. The judicious use of data structures within the global data area could avoid such problems but it would require specific assignment statements to the appropriate record structure. It should be noted that all modules that access the global data area should be recompiled when changes are made to the global database because static memory areas may be affected by changes in the declarations. There is a clear need for Configuration Management support to avoid mismatches in object code.

Procedural Parameters

Data passing via parameters has the advantage that code can be reused and that modules are only allowed access to the data required to perform their assigned tasks and that the results produced by static analysers are more understandable and tractable. This technique is considered to be good systems engineering practice, but if used excessively can result in performance degradation.

3.5.5 Language Considerations

The language used to develop the system is an important decision, i.e. some languages support global data areas and modularisation better than others. Advice on the selection of languages recommended for safety-related systems may be found in (IEC 1508 draft standard [6] and *The Choices of Computer Languages for use in Safety Critical Systems* [25]). The IEC 1508 draft standard [6] also recommends that languages be problem oriented rather than machine oriented.

Code should always be well commented, however some languages are more readable and thus easier for an independent reviewer to understand. The call structure imposed by the implementation language should be clear. It is possible to implement an algorithm in many different notations while remaining functionally equivalent, some languages make this easier to do than others. There is a need in such cases for coding standards to be developed and enforced. In some cases a language sensitive editor may be available that prevents the programmer from entering undesirable language constructs. Such editors tend not to enforce the implementation of algorithms in a favoured manner. In such cases it is necessary to perform code audits by humans.

The language employed will influence the manner in which logic may be implemented. There are three contributing features of a language that affect the ease of implementation of logic: the data types supported; the operations possible on supported data types; and the control constructs provided by the language.

Difficulties arise in PLCs in that the programming languages' 'ladder logic' tends not to support the use of meaningful names for variables. This makes understanding of the function of the code difficult during review.

It should be possible using structured design to have clearly identifiable modules, which separate control logic from algorithms. However, PLC languages tend not to support structured programming. Only one module or 'ladder' can exist, 'structure' is imposed by using flags to enable blocks within the ladder to be processed. Such blocks must be positioned correctly in relation to the flag manipulation rungs on the ladder to avoid incorrect operation. High-level languages (such as Ada and C++) support the use of meaningful names for variables and provide for well structured, readable code.

3.5.6 Software Variants

Multiple instances of systems may require variants of a common or core software. The variations could be introduced by the number and types of I/O points used, or by different display facilities required.

A system may be made configurable if it is to be used on a number of slightly different sites, but each has a common operating philosophy.

If a system is configurable to create such variability it will be more complex than its non-configurable version for a given application. This will be due not least to the code introduced to search for data held by the system configuration database. This code adds to the basic path measure of complexity and will thus affect the test coverage achievable. Even once configured, a configurable system will be more complex that its non-configurable form, due to the use of variables imported from configuration files rather than constants.

One can argue that if a system is configurable then it should always be tested in the configured form for the site. With the reuse of the software over a number of sites it may become economic to produce a configurable plant simulator that matches the PES configuration to the plant. The PES can then be driven through a series of standard tests to ensure that the particular configuration has not created an unexpected change in function.

3.6 Guidance

- Effort should be made to include 'test points' in the design, which may be observed externally to the processes under test.

- In time-critical systems where additional software instrumentation cannot be tolerated the use of hardware monitors should be considered.

- In comparing four classes of system for testability, the simplest implementation types emerged with the most favourable score.

- Systems should be simple if they are to be testable. By selecting simple operating systems, and designing systems that recognise this simplicity, unseen complexity will be avoided in safety-related systems.

- The operating system used in safety-related systems should be configured so that timing can be guaranteed. This may involve segregation into separate systems that provide dedicated functions and which have well-established time deterministic behaviour.

- Drafts of future standards for system development indicate that software covered by the standards is likely to include operating systems, compilers and test routines as well as application programs. The requirements for testing such support software will be generally at the same level as the application software.

Chapter 4
Testing of Timing Aspects

Timing is one of the most critical aspects of a 'hard' real-time control system. Achieving critical timing constraints can be very difficult and presents a major challenge to the system designer. As any errors in the specification of timing requirements have an effect from the system level through to the software, rectifying errors in requirements can be expensive both in time and cost. Ensuring that timing requirements are clearly specified and validated early in the design phase brings important benefits.

There are two main timing issues; the timing constraints placed on the system from the system requirements and implementing a scheduling system to meet these requirements and the calculation of software timing characteristics.

4.1 Introduction

Timing constraints define response time requirements for the software and/or its environment. Timing constraints for non-real-time systems are very simple to express, and all that is required is a few short sentences in English. Timing constraints for safety-related 'hard' real-time systems are more complex to express and require a higher degree of unambiguity in their specification. There are essentially two categories of timing constraints for real-time systems:

- performance constraints that set limits on the response time of a system
- behavioural constraints that impose constraints on the rates at which the environment can apply stimuli to the system.

One way of categorising timing constraints is by stating to what they apply, that is, the system or environment.

They can be further classified by three types of temporal restriction:

Maximum   No more than a set amount of time may elapse between the occurrence of one event and the occurrence of another.

Minimum   No less than a set amount of time may elapse between two events.

Duration   An event must occur for a set amount of time.

An event is defined as being either a stimulus of the system from its environment, or an externally observable response that the system makes to or on its environment.

There are also other types of constraints that do not explicitly specify a timing constraint but essentially have timing constraints embedded within them. Consider the following:

The combustion area temperature should never be greater than 500°C

This may not at first sight seem to be a timing constraint. But if we consider the following:

- The system takes a finite amount of time between reading an input and performing the necessary action.
- As the software runs in the digital domain the required parameters are only sampled at discrete times. To ensure the constraint is met the designer must reason about the rate of change of temperature. This is obviously time dependent.

The system designer has to ensure that the software will run fast enough on the platform to ensure that the above criteria are satisfied.

4.2 Correctness of Timing Requirements

Once the requirement specification documents are finalised how does the software development team know that all the requirements, especially timing, are correct? It is advantageous to verify the requirements as the system design takes shape. This promotes the idea of both reducing the overall cost of producing the final version of software and also builds confidence on the system under development during the development lifecycle.

The following questions need to be considered when reviewing the requirements specification:

- Are the requirements correct?
- Are they consistent with each other? i.e. do we have contradictory requirements?
- Are there any superfluous requirements specified due to pessimism by the author of the requirements document?
- Are they specified too rigidly? Could they be defined more flexibly?
- Have we a full set of requirements?
- Are they achievable with the proposed system architecture?

Other problems can also occur when the final testing against the timing requirements is performed. Historically, the verification of timing requirements is carried out after the code has been written. The verification method normally includes running the software in the target hardware, and using some form of hardware based tool, for example, a logic analyser, to monitor the software as it is being executed. This can be inefficient for the following reasons:

- If a constraint is not met, at best the code is redesigned. This could mean that a compromise is made to the algorithm to meet the timing constraint. Also, if after a redesign of the algorithm, the constraints are still not met, the system developers sometimes revert to writing various parts of the system in assembly language to speed up the execution time. This often leads to long term maintenance costs and poor quality systems.

- When considering hard real-time (safety-related) systems the question arises as to whether or not worst-case execution times should be considered rather than just the ad hoc running of the system. Considering worst-case execution times for all items of code may be unrealistic but when considering safety-related systems it should ensure that the system will meet its timing constraints under any system load.

- As testing is naturally later in the software development lifecycle and can account for nearly 50% of the development cost by the time faults are rectified, any changes in system design can be very costly in time and money.

Currently, most safety-related hard real-time systems are designed using methods that do not provide any guarantee that any timing constraints will be met. As a result, timing errors where computations miss their deadlines, are the most unpredictable, most persistent and most difficult to detect. Also at present there is no formalised method for specifying timing constraints. Existing requirement documentation is normally written in English with any timing requirements either expressed in English or at best presented in some graphical or mathematical format. Specifying timing requirements in natural language leaves a great deal to be desired because:

- Natural language relies on the shared linguistics of the author and the reader of the requirements. Requirement authors assume that the terms used in the specification mean the same to the readers. This can be a dangerous assumption due to the inherent ambiguity of natural language.

- Unstructured paragraphs of natural language are unable to express the overall functional architecture of the system in a clear and concise way.

- Natural language is over flexible in that it allows functionally related requirements to be specified in completely different ways. It then falls to the reader to identify and group related requirements.

- Requirements are not partitioned adequately. The effect of any changes can only be determined by examining each and every requirement rather than a group.

An alternative to natural language is to use formal specification languages like Timed CSP and Petri Nets. The advantages are as follows:

- The development using a formal specification often provides better understanding of the software requirements.
- Software tools if made available could aid with the development, understanding and debugging of requirements.
- It may be possible to animate the formal specification.
- As formal software specifications are often mathematical models they can readily be studied and analysed using mathematical methods.
- Formal specifications may be used to help guide the tester in choosing the appropriate test cases.

Utilising formal specification languages such as Timed CSP can increase the confidence in any timing specifications placed on the software. Timed Petri Nets are also useful for modelling the behaviour of small real-time safety-related systems and are highly recommended for use in small safety-related projects.

4.3 Scheduling Issues

The scheduler's task is to select which program may execute and for how long. At present there are many different techniques that can be used to implement a real-time scheduling system. The choice of technique is normally heavily influenced by the final application.

There are two major approaches to scheduling algorithms; static and dynamic. Static algorithms fix schedules before run-time and dynamic algorithms determine schedules during run-time. Schedulers may also be pre-emptive or non-pre-emptive. That is, when the real-time clock interrupts, a pre-emptive scheduler examines the priority of all tasks currently runable, and the task with the highest priority will be allowed to execute. A non-pre-emptive scheduler allows a task to execute to completion even if a higher priority task wishes to execute. Much work is being done on pre-emptive scheduling algorithms, such as on the rate monotonic algorithm, to prove that the behaviour can be predicted statically before execution.

It is well known that real-time safety-related systems should be deterministic, so historically, timing integrity is achieved by using a deterministic static cyclic scheduling system. This is because a system that uses a dynamic or pre-emptive scheduling system is inherently non-deterministic. Much research into priority pre-emptive and dynamic schedulers is currently being undertaken in academia. This includes mathematical proofs to predict that before implementation the proposed scheduler can meet all timing constraints, even under worst-case conditions. Unfortunately, at present, these theories are moderately new and the risk of using dynamic schedulers on a safety-related system outweighs the advantages to be gained by using them.

The choice of scheduling strategy greatly affects the type and amount of testing required of the final system. The type of scheduling strategy is affected by the nature of the system under consideration.

The following paragraphs outline the characteristics to be considered when selecting the appropriate scheduling algorithm.

4.3.1 System Attributes

Hard real-time systems typically consist of a large number of processes all with rigid timing constraints combined with high processor utilisation. Tasks within hard real-time systems can be thought of as a set of co-operating sequential processes, where a process consists of a set of operations to be performed in a prescribed order. Processes with deadlines critical to the correct functioning of the software are described as having hard deadlines. Processes that can occasionally miss their deadlines are described as having soft deadlines.

Two distinct process types can exist in real-time systems:

- *Periodic* processes execute repeatedly at a fixed iteration rate. A typical use of a periodic process would be to read sensor data at regular time intervals.

- *Sporadic* processes respond to an external or internal event. A typical use of a sporadic process would be to respond to an input switch.

Within hard real-time systems the majority of processes will be periodic, with very few, if any, sporadic processes. This may be because hard real-time systems are concerned with simulating continuous behaviour and thus require many computationally intensive periodic processes. Also, data on external behaviour can be buffered and processed by a periodic task in the required time. This alleviates the use of several levels of interrupts that are considered undesirable for hard real-time systems as the elapsed time spent in the interrupted processes is unpredictable.

In order to allocate processor time to each process, some form of allocation algorithm or schedule is required. Two distinct approaches to scheduling exist:

- Run-time scheduling computes the schedule on-line as the processes arrive. This is known as dynamic scheduling. The advantage is that the schedule can be adapted to suit changes in the environment during run-time. The disadvantage is that the scheduler itself requires processor time to compute the schedule and the scheduler is deemed non-deterministic.

- Pre-run-time (or static) scheduling uses schedules computed off-line with a different schedule for each mode of operation. A small run-time scheduler is then used to select the correct schedule in response to external or internal stimulation. The advantages are low processor overhead and determinism. The disadvantage is that sporadic processes are difficult to schedule.

For hard real-time systems the objective of the scheduler is to ensure that all the processes meet their deadlines. Therefore, predictability of the system's behaviour is the most important concern, and is why static scheduling is almost always used for hard real-time systems, even at the expense of treating what would otherwise be sporadic processes as periodic, as described below.

When considering scheduling algorithms the types of processes in the system affect the choice of algorithm.

- Static – The majority of scheduling algorithms are designed to schedule a static set of processes; for which start times, deadlines, and computational times for each process are known.

- Periodic – Some algorithms are concerned explicitly with the scheduling of a set of periodic processes. A static priority scheduling algorithm can be used to schedule periodic tasks if the set of periodic tasks can be scheduled within a time period between zero and the least common multiple of the process periods, with the schedule being repeated to form a major cycle.

- Sporadic – On-line scheduling algorithms can be used to schedule tasks with unknown start times. However, unless the system is simple, no guarantee as to whether the system is scheduleable can be made before execution. If the worse-case delay bounds are known, the sporadic processes may be converted into equivalent periodic processes and a static scheduling algorithm can be used to compute a feasible schedule.

- Both periodic and sporadic – The above method can also be used to guarantee in advance that both periodic and sporadic processes will meet their deadlines. This would mean translating sporadic processes into equivalent new periodic processes, and then using static scheduling algorithms to compute a feasible schedule for both the original and new periodic processes within a time interval between zero and the least common multiple.

4.3.2 Number of Processors

The number of processors in the system architecture is a deciding factor in the scheduling algorithm selection, and the following considerations apply:

Single processor Most current scheduling algorithms are designed for uniprocessor systems and generally cannot be applied to multiprocessor systems.

Two processors A few scheduling algorithms are designed specifically for use with dual processors.

N pre-assigned processors Normally the scheduling algorithms pre-assign each process to a CPU and during execution the process always executes on its pre-assigned CPU.

N processors Generally algorithms allow CPUs to be allocated to individual processes on-line as or when required according to the current system state.

4.4 Scheduling Strategies

4.4.1 Static Priority-Driven Schema

Each process is assigned a priority before execution and at each instant during run-time the resources are allocated to the process with the highest priority. The schema can be either pre-emptive or non-pre-emptive. Static priority schedules only provide a limited subset of schedules for a given number of processes, which restricts the ability of the algorithm to satisfy all timing constraints. Also, the smaller the number of schedules produced by the algorithm, the lower the level of processor utilisation. If a 'priority pre-emptive' based scheduler is used, the actual execution order of the processes may be different to the original priorities, this is due to process priorities being changed to suit each particular circumstance. For example, if a program was to enter a critical section it may block every other process from gaining system resources until it leaves its critical section. This makes the verification of all timing and resource constraints before execution very difficult. Research into the prediction of the behaviour of static pre-emptive algorithms before execution, such as the rate monotonic algorithm, is being undertaken.

If any type of pre-emptive or dynamic scheduling algorithm has been used, then the only way at present to ensure all timing constraints will be met at run-time is by testing. The fundamental flaw in this practice is that testing only shows the presence, not the absence, of problems. Also, because satisfying timing constraints in hard real-time is inherently deterministic, stochastic simulation can only show the average timing behaviour of the system and does not guarantee timing behaviour under worst-case conditions.

4.4.2 Pre Run-Time Schedules

When considering hard real-time systems, it is assumed that the system consists of a large number of independent processes with stringent timing constraints and very high processor utilisation and with very little spare CPU capacity. If the system is simple, has very few processes and large CPU capacity, then the problem is simple to solve. Unfortunately this is rarely the case. When choosing a scheduling algorithm to be implemented the following must be considered.

The Satisfaction of All Timing Constraints The idea of using mathematical techniques to solve the problem of calculating whether a feasible schedule exists. The problem of finding a feasible schedule mathematically is difficult; even the simple static cyclic schedule problem is believed to be NP-complete. (See Section 4.4.3)

The Control of the Execution Order Of Processes Large processes are sometimes decomposed into smaller more manageable processes. These processes therefore need to be executed in some order relative to each other. This imposes further time constraints as the individual smaller processes would have to be executed in the correct order within a finite time period.

Historically, the scheduling of safety-related hard real-time systems was done by hand. This inevitably leads to 'spaghetti' code that is hard to verify and maintain. Such systems are also very time consuming to design and quite often will not find a feasible schedule.

The use of appropriate scheduling algorithms can facilitate the task of pre-run-time scheduling. Their application quickly reveals if a feasible schedule exists. Furthermore, if schedules can be computed automatically then new schedules can be computed when the system is modified.

The amount and type of final testing required of the final system depends upon the scheduling strategy chosen. Dynamic strategies obviously require more verification than that of static strategies as final system behaviour is considerably harder to predict before run-time. Static strategies, such as cyclic schedulers, can be calculated using various optimisation techniques such as simulated annealing. Using simpler strategies reduces the level of testing required as the schedules are simpler to calculate and therefore more reliable.

The normal choice for safety-related systems is to use some form of static cyclic scheduler. One way of ensuring that the system will meet its timing requirements is to use some tool to determine whether a static cyclic schedule is feasible for the software.

4.4.3 The Static Cyclic Scheduler

The overall structure of a cyclic schedule is determined by its major and minor cycle times. The major time is the interval at which the entire schedule is repeated. The minor time represents intervals within the schedule at which the start of execution of a process may be synchronised. Instances of process execution are assigned to minor cycles in order to satisfy the process iteration, thread ordering and end-to-end deadline requirements. Clearly, processes must not be inserted into minor cycles such that, in the worst case, more time is consumed by process execution within the minor cycle than is available (i.e. frame overrun).

Finding whether or not a solution to the problem exists is non-trivial. The cyclic schedule problem is readily compared with a problem known as the bin-packing problem. Consider the problem of attempting to fit a number of different sized objects into a fixed number of bins. The problem of calculating whether the objects can be placed in the bins belongs to a set of problems known as NP-complete.

The bin-packing algorithm can be applied to the scheduling problem by equating the minor cycle to a bin and equating its time to its size, for example 25 ms. The number of minor cycles can then be the number of bins. The execution time of each software process can be equated to an object size. See Figure 4-1.

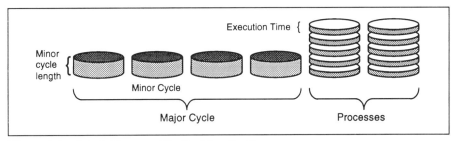

Figure 4-1: Scheduling as a Bin-Packing Problem

The problem is complicated further because a process can appear in more than one minor cycle. For example, given a 100 ms major cycle and a 25 ms minor cycle, a process with a 25 ms iteration rate would need to execute in every minor cycle. Also, any end-to-end timing requirements need to be considered.

4.4.4 Approaches to the Static Cycling Scheduling Problem

There are various search techniques that can be applied to the calculation of a static cyclic scheduler.

- Brute-force searching involves the evaluation of each possible solution in order to find a feasible solution. Once a feasible solution is found the search can also be continued until the best solution is found.
- Heuristic algorithms use knowledge of the system to direct the generation of solutions to be tested with the sole aim of reducing the number of solutions to be tested. For example, the algorithm starts with an empty schedule (i.e. one with no process executions within it), and inserts threads one at a time. Firstly, threads are inserted in order of end-to-end deadline, with the shortest deadline first; this is based on the premise that threads with short deadlines will be harder to manage than those with longer deadlines. Secondly, process instances are inserted into the schedule as late as possible within the minor cycles within which they appear; ideally, each time one is inserted, it is placed as the last process instance in the minor cycle. However, a process instance is always inserted earlier in a minor cycle than any process instances already in that minor cycle that occur after it in its own thread. Where possible, the process instances for each thread are inserted into the schedule in the least loaded minor cycle. However, this is not always possible as if some process within the thread has already been inserted, then other processes are inserted into the same minor cycles.
- Simulated annealing is a technique related to the annealing of materials, i.e. a slowly cooled material will settle into a regular structure. The method uses knowledge of the relative value of possible solutions to calculate the quantitative measure of value of a proposed solution. This

enables the algorithm to determine whether a proposed solution is better or worse than the current solution.
- Stochastic evolution is essentially a modification of the simulated annealing algorithm which attempts to reach solutions more quickly, or better solutions in the same time.

4.4.5 Simulated Annealing

Of all the search techniques, simulated annealing has been shown to be the most useful for generating cyclic schedulers. Simulated annealing requires two functions that depend on the problem in hand, and a driving algorithm, which is essentially problem independent. The first function is the neighbour function; given a particular solution, the neighbour function generates a different solution, related to but randomly differing from the first. The second function, the energy function, takes a solution and calculates a numeric value that measures the usefulness of the solution, a better solution having a lower energy than a worse solution. The application of the energy function to the space of all possible solutions can be viewed as a landscape, where valleys represent better solutions, and ridges represent worse solutions. A particular altitude corresponds to the division between feasible and non-feasible solutions; the surface below the altitude encompasses the feasible solutions. The search for a feasible solution corresponds to the search for a point on the surface below this altitude, and for the best solution to finding the lowest point on the surface.

A driving algorithm generally functions as follows:

- An initial solution is chosen at random, and an initial temperature selected.
- A new solution is generated using the neighbour function, and the values derived from the energy function for each solution are compared. If the new solution is better (has a lower energy), then it is used and the older solution discarded. If it has a higher energy, then it is still used with a probability that is exponentially related to the energy difference and the temperature.

To mimic the annealing process, the temperature is gradually reduced, reducing the probability of accepting new solutions that have higher energies. This process continues until a feasible solution is found. The annealing process need not stop there; it may continue in order to find a 'better' feasible solution.

An initial temperature is chosen such that a large fraction of new solutions are accepted. This ensures there is a good chance of moving from one valley on the surface to another, without being blocked by intervening ridges. Hence, the search is prevented from becoming stuck in a shallow valley (called a local minimum). The algorithm has to take into account the following when calculating a feasible schedule:

Iteration rates   Each process must iterate at its specified rate. For each process, the energy function notes the first minor cycle in which a process instance runs. It then finds the subsequent process instances, and adds to the energy function if that instance is not in the required minor cycle.

End-to-End Deadline Times Each thread must execute within the specified end-to-end deadline. That is, the time from the start of execution of each execution instance of the first process in the thread, to the end of execution of the corresponding instance of the last process in the thread, must not exceed the end-to-end deadline. For each thread, and for each execution instance of the first process in the thread, the energy function scans through the cyclic schedule to locate execution instances of successive processes in the thread. Note that this scan ensures a correct flow through the process instances, so that a process instance will be ignored if it is disordered with respect to the thread. If the end-to-end deadline is exceeded, then a contribution is added to the energy value.

Minor Cycle Overload Clearly, the sum of the worst-case execution times of the process instances in a minor cycle must not exceed the minor cycle time. Each minor cycle is checked, and the energy value increased for each overloaded minor cycle.

Minimum Time Gap in Input-Output Processes As threads are checked for end-to-end deadlines, the gaps between execution of processes within the thread are checked. If any are less than the specified minimum, then the energy value is increased.

Penalty and Benefit Although any feasible solution is in principal acceptable, some solutions are better than others: that is, they will have lower energy values. Ideally, the algorithm should find the best solution of all. However, although this is computationally impossible, it is nevertheless desirable to find a good feasible solution. However, any infeasible solution is obviously worse than any feasible solution. For this reason, the energy function actually generates two values, the penalty and benefit values. The former is the value described above. The benefit value is calculated where a requirement is met. For instance, if a thread end-to-end deadline is met, then the benefit value is adjusted by the time by which the actual end-to-end time is less than the deadline. Similarly, if no minor cycle is overloaded, then an even spread of free time is beneficial; and provided that processes iterate in suitable minor cycles, a regular execution interval (in actual time) is beneficial.

It has been demonstrated that such a tool can be produced and is capable of calculating feasible schedules for a large number (greater than 100) of individual processes. The tool can be utilised during the software design stage to gain confidence in whether or not an acceptable scheduling algorithm can be found. Estimates of execution times can be used but obviously any such tool is highly dependant on the calculation of the worst execution times of the individual processes.

4.5 Calculating Worst-Case Execution Times

The execution times of individual software packages need to be known to perform any static pre-run-time scheduling analysis. Traditionally, measurements have been taken of the run-time execution behaviour of the program but this produces accurate average case timing behaviour but poor worst-case timing.

Many attempts have been made to analyse the timing behaviour of programs but all seem to encounter the same set of basic problems:

- Some programming language constructs demand an unbounded length of time.
- Source code does not contain sufficient information about paths through the code.
- Modern computer architecture encompasses many non-deterministic features that are implemented to improve overall speed.
- Some techniques produce worst-case execution times that are pessimistic. The technique should produce accurate worst-case values that will be encountered at run-time rather than the longest execution path through the code.

4.5.1 Language Issues

Commonly used high-level programming languages are not designed with timing analysis in mind. However, some research languages have been designed with timing in mind (for example Real-Time Euclid) but none have found widespread use. Popular languages such as Ada, C++ and Pascal include features that make accurate timing analysis infeasible. To make such languages amenable to timing analysis, language features need to be excluded to form a language subset. Subsets specifically aimed at improving timing calculations generally enforce the following restrictions:

- The use of the GOTO statement is prohibited. The use of the GOTO statement produces a program that has no analysable flow of control.
- Recursion is prohibited. Recursion is particularly difficult to handle and can lead to unpredictable timing.
- Loops need to be bounded in some way. This merely requires that all loops should terminate within a known fixed time or a fixed number of iterations.
- Program units, such as functions and procedures, need to be designed with a single entry and single exit point. This simplifies their analysis and allows them to analysed in isolation.
- Dynamic memory allocation should not be allowed. This is because current dynamic memory allocation schemes have unpredictable timing behaviour.

The subset above is very restrictive and the ideas behind calculating worst-case execution times are still in their infancy.

4.5.2 Simple Example

To illustrate the basic idea behind worst-case execution timing calculation, a small fragment of high-level code, in some suitably constrained language, is used. Consider the following 'if' statement:

```
if (expression) then
    statement1;
else
    statement2;
end if;
```

A common technique for all timing analysis is to compose a basic-block structure of the code using control flow analysis. Control flow analysis dominates most static analysis techniques and is used to model the flow of control through a program. A basic block is defined as a sequence of instructions that has a single entry point and if one instruction in the block is executed then all are. The above example consists of three basic blocks: expression, statement1 and statement2. These basic blocks could be further broken down into more basic blocks. The control flow for the above code is shown diagramatically in Figure 4-2.

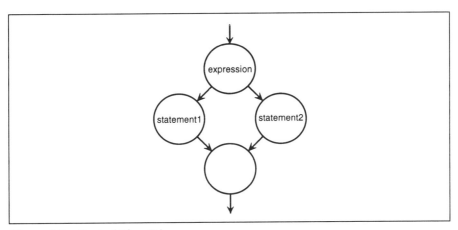

Figure 4-2 : Control Flow Diagram

The worst-case execution time (WCET) for this fragment would be given by:

WCET(if-statement) = WCET(expression) +
max (WCET(statement1), WCET(statement2))

As statement1 and statement2 could consist of further sequences of basic blocks, the calculation needs to be performed in a 'bottom up' manner. This example highlights a basic problem with the calculation of WCET. If at runtime the 'expression' part of the *if-statement* always equates to FALSE, then one path could be eliminated. The argument also applies to other language constructs, particularly any looping constructs that often use variables as limits. Obviously, any simple WCET tool would calculate a time, but this would be extremely pessimistic and this needs to be taken into account when choosing the target processor. To assist in solving this problem, the code needs to be annotated with extra information; for example, the maximum number of loop iterations.

4.5.3 Low Level Timing Analysis

When considering the high-level language we assumed that timing information can be calculated for each basic block of code. A simple approach is to calculate the execution times of all assembly code instructions used to implement each basic block of code. For example, using Motorola MC 68000 assembler code:

Code		Clock cycles
MOVEM.L	A1-A2,-(SP)	8+8n
MOVE.W	#4,CCR	6
CLR.L	D0	6
MOVEB	(A1)+,D0	8
ASL.L	#8,D0	8+2n
ADD.B	(A1)+,D0	8
MOVE.W	D0,(A2)	8
MOVEM.L	+(SP),A1-A2	12+8n
Total		138 cycles

Unfortunately, the task of calculating WCETs is not as simple as this because certain complications arise due to:
- architecture and memory constraints
- source of instruction stream
- hidden or unexpected control flows
- processor caches and pipelining
- input/output.

4.5.4 Architecture

Target system architecture has a major impact on the system timing and must be considered. For example, consider a simple system that consists of a single CPU, some memory, I/O devices and a single communication bus to connect the system components together.

The CPU has a specified speed (generally specified in MHz). From this we can calculate the time it takes the processor to complete a single execution cycle.

In order to perform computations, information will be required from the system memory. When obtaining data from a required memory location (ignoring the use of cache) the following must be considered:

- Memory access time – the time taken between presenting the required address to the valid data appearing on the bus.
- Memory cycle time – the minimum time allowed between consecutive interrogations of the memory.
- CPU access period – the interval after which the CPU will expect valid data.

If the memory access time is longer than the CPU access period then the CPU must wait until the memory supplies the valid data. This inevitably means that the execution time will increase due to the overhead of the processor having to wait for the memory. This should be taken into consideration when calculating WCET.

The type of memory used for implementation can also affect WCET calculations. Two types of memory have become popular during implementation: dynamic and static. A dynamic memory cell consists of a capacitor and a transistor. As the capacitors leak charge they must be refreshed periodically, which consumes time on the memory bus. This combined with the longer cycle times degrades system performance and should be taken into consideration. Static memory does not suffer from this problem and normally has an access time equal to the cycle time.

When considering memory we must also consider bus width. To access a 16-bit word across a 16-bit data bus would only require one cycle, but to access the same 16-bit word across an 8-bit bus would require two cycles. This would increase execution times.

Modern processors often include onboard cache and instruction pipelining. For safety-related applications these features are generally switched off (if possible) due to their non-determinism, so that simple timing analysis can be performed without the fear of further pessimism being introduced. For non-safety-related applications the mechanisms may be required to improve system performance. Therefore, performing simple timing analysis introduces pessimism into the calculations. The use of performance enhancing architecture and its effect on timing estimation is still a topic of research.

Other system architectures prove difficult to analyse, these include the use of numerous execution units within the system, for example floating-point co-processors and dedicated graphics processors. Modelling a system that includes distributed processors is extremely difficult.

The final source code needs to be available to analyse how long each basic block of code will take to execute. This enables timing information for each assembly code instruction, which is normally supplied by the processor manufacturer, to be utilised.

Two techniques are possible:

- *instruction prediction* – attempting to predict the instructions that will be generated for a piece of code.
- *object code reading* – taking the object code as produced by the compiler.

Unfortunately both techniques are problematic due to the action of the compiler. High-level language basic block structures may be lost from the object code due to optimisation by the compiler; this is especially true for Ada compilers.

4.5.5 Hidden Control Flows

Due to the nature of some high-level language constructs, compilers can introduce additional control flows. For example, when equating two array segments:

```
FASTCARS(1..10) = GENERICCARS(100..109)
```

the compiler could introduce a loop consisting of ten MOVE instructions. The amount of compiler intervention is dependent on the compiler itself and the language itself. For simpler languages, such as C, a straightforward mapping to object code is possible, but for complex languages like Ada, it is not. One approach in overcoming this is to analyse the assembly code produced from the high-level code. Unfortunately, attempting to associate the control flow of the high-level language with assembly code is extremely difficult.

4.5.6 Input/Output (I/O)

Input/output characteristics are extremely pertinent to real-time systems because they always involve some communication with their associated environment. Three basic methods exist for dealing with I/O.

CPU initiated I/O The device registers are mapped onto specific areas of memory allocated for the device, or special instructions are used to transfer data. This can either be unconditional, whereby data is always available when required or conditional whereby the CPU must wait for the device to be ready before any transfer can occur.

Device initiated I/O The devices inform the CPU when they are ready to transfer data by raising an interrupt. The interrupt can then be handled by context switching the current program out, performing the I/O and switching the original program back in.

Direct memory access. Dedicated hardware is used to transfer data directly to main memory via the bus by cycle stealing from the CPU.

CPU initiated I/O has no impact on WCET analysis as the method of transfer uses specific statements with known execution times, or the program simply loops for a known period of time, i.e. polling. Device initiated I/O uses interrupts as the control mechanism so the problem becomes a scheduling problem and does not effect the WCET calculations. DMA is similar as it does not effect the WCET of the individual program units. However, as DMA interrupts the CPU in a non-deterministic manner, timing analysis of the whole system becomes difficult.

4.6 Guidance

- Dynamic testing methods can be used to test non-functional properties such as timing but adequate testing to achieve the required level of confidence is extremely expensive. An effective alternative is to use static analysis techniques such as pre-run-time scheduling to complement the dynamic testing.

- Timing analysis should be considered through the development lifecycle from requirement specification to final validation testing. Various formal languages (such as Timed CSP) are emerging that consider timing to be an integral part of the system requirements which need to be verified before software should be written.

- Some modern CASE tools incorporate features that allow system models to be created and evaluated with respect to timing. This technique is useful for small systems with small numbers of processes but becomes difficult with larger systems with a large number of co-operating sequential processes.

- As code metrics become available, application specific tools can be used to determine whether the final system can be scheduled.

- Pre-run-time schedule calculation tools are heavily dependant or the accurate calculation of worst-case execution times (WCET). The calculation of accurate WCET for any high-level language is non-trivial. A language subset needs to be specified to remove all language features that are non-deterministic.

- Currently there are two options for the calculation of WCET:
  - Predict which low level instructions will be used to implement the basic-block structures of the code and then calculate WCET from information supplied by the processor manufacturer.
  - Use the object code produced by the compilers and determine the relationship to the high-level code basic-block structures.

- Hardware implementations also need to be considered in the calculation of the WCET, because the type of memory, bus, etc. can all affect the calculation.

- Current research seems to be concentrating on the run-time/pre-emptive scheduling options. However, the ability to produce and verify a deterministic scheduler statically is crucial for hard real-time systems.

Chapter 5
The Test Environment

This chapter concentrates on the requirements for a generic set of test tools together with their associated quality requirements and lists some tools which are commercially available. A number of widely known tools which are commercially available and fit into the proposed toolset are reviewed. Advice on the selection of tools is given. The other elements of a test environment are stubs and harnesses and plant simulators. Plant simulators are covered in Chapter 6 but stubs and harnesses are mainly project specific and so are not considered in detail within this book.

Note that this chapter contains opinions based upon the investigations carried out within the CONTESSE project and not on a comprehensive review of testing tools. The inclusion or exclusion of any tool does not imply any ranking amongst all available tools. A project specific investigation into the suitability of tools should be carried out if an organisation is considering the use of tools to assist in the testing of safety-related systems.

5.1   Introduction

The main objective when introducing test tools is to increase development productivity (and hence concentrate time on safety-related activities) and software quality (with the obvious link to safety). A wide range of high quality commercial tools exist and it is the responsibility of the project staff to select those tools that will provide a cost benefit and contribute to safety within the context of the specific project.

Although the actual choice of tools is project specific, it is possible to describe a generic set of tools which would support the test activities related to the safety case. Taken together with an understanding of the safety and quality requirements that must be met by the chosen test tools, this set provides a useful aid to the project specific selection of a set of real tools.

In addition to proposing a generic tool set and selection criteria, this chapter also reviews three commercially available tools for their suitability for safety-related applications.

Note: The assessment of commercially available tools was carried out in 1995. Some or all of the drawbacks highlighted may have been overcome by the release of new versions of the tools. The suppliers of the tools should be contacted to obtain the most up to date description of their capabilities and the tools reassessed for their suitability for use on any project.

5.2 Test Activities Related to the Development of a Safety Case

The test related development activities that contribute to the production of evidence for the safety case may be correlated with the contents of the safety case as outlined in Table 5-1 below. The development activities can be further summarised as:

- identification of requirements (including risk reduction requirements (RRR))
- development of the Specifications (which maintain traceability to requirements, including RRR)
- specification of tests to demonstrate compliance with requirements.

Table 5-1 : Test Related Development Activities which Contribute to Safety Case Contents

Activity	Safety Case Content
Identification of Hazards and HAZOP	Results of hazard and risk analysis
Tracing of hazards to system specifications which demand protection against hazards	Details of risk reduction techniques employed
Defining sets of specifications (functional, design, module)	Results of design analysis showing that the system design meets all the required safety targets
Identifying a test regime commensurate with the required integrity of the system	Test strategy linked to design analysis
On completion of appropriate development stages, the execution of tests and analysis of results	Results of all verification and validation activities

5.2.1 Identification of Hazards

This aspect of the safety lifecycle impinges on the PES development lifecycle at its earliest stage when the requirements are first being stated. There are a number of structured methods used in risk assessment, and in some cases there are tools available to support capture of data generated by the identification of hazards. As part of the conceptual design of plant it is common for a simulation, based on the physics of the process, to be carried out. Such simulations can be used to investigate plant behaviour under closed loop conditions, to assess required response times and, if desired, can be extended to include plant failure conditions. Thus plant simulation can be a useful tool to support identification of hazards.

5.2.2 Tracing of Hazards

Given that the identification of hazards takes place, a set of requirements will be developed which indicate to the PES developer the behaviour required of the PES to avoid hazardous situations. Therefore, within the PES development lifecycle the aim is to maintain traceability of each step in the development to these requirements.

Throughout the development of the PES further analysis of the system is required to demonstrate that the integrity of the system specification has been maintained as design detail is added.

5.2.3 Defining Sets of Specifications

As development progresses and more detail is added it becomes more difficult to assimilate all information and check on the satisfactory behaviour of the system. Structured decomposition methods are used to reduce the complexity of individual items. However, the overall complexity of the system may still be high. Tool support exists for a number of analysis and design methods. Common techniques include structured analysis with real time extensions, object oriented analysis and design, and structured design. Traceability between requirements and design needs to be maintained.

5.2.4 Identifying a Test Regime

As part of the safety justification of a system, evidence that adequate testing has been carried out must be provided. However, in the first instance, a set of adequate tests needs to be determined. The test regime assessment model, discussed in Chapter 9, offers a structured means of developing an argument for a set of tests. This model is based on the advice offered in a number of standards and guidelines related to the development of safety-related systems containing software, that is considered to be commensurate with the system's required level of integrity. The model consists of a number of assessments of test methods and their applicability. The outcome of these assessments is a test regime score that indicates the likely compliance of the proposed test regime with the various standards and guidelines. As these guidelines provide some flexibility in choosing test methods there needs to be some way of selecting a set of tests. A tool that would allow the assessment of the sensitivity of the Test Regime score to choices made is considered to be useful.

5.2.5 Execution of Tests and Analysis of Results

No matter what type of testing is to be performed, some form of test harness will be required. The form of the test harness could range from a simple stub and driver code harness to a full plant simulator interacting with the system under test through its actual plant interfaces. The problem of analysis of results should not be overlooked as it is probably by far the hardest part of testing; the capture of test data can be trivial by comparison. When white box testing is to be performed a number of measures can be employed to enforce a systematic approach to stimulation of code sections. If such an approach is to be used, tools that identify code paths to be tested and monitor execution of modules for coverage of the identified paths can assist in the management of tests. Using just some of the white box coverage criteria can lead to a large number of tests. In such cases a means of managing all the test data and results files, test harness and analysis code is useful.

5.3 A Generic Test Toolset

A generic set of tools to support the test activities described above are:

Tool A A simulation tool that permits functional requirements to be modelled and executed and allows for systematic testing to be applied much earlier in the development lifecycle.

Tool B A tool which maintains links between requirements test cases and system elements to be tested and can report on these links and the status of each of the objects related.

Tool C A tool that supports calculation of test regime compatibility with required integrity.

Tool D A complexity analysis tool for control of development of code.

Tool E A code coverage analysis tool based on actual execution of code.

Tool F A test case management system to create and monitor and report on test case execution.

Tool G A simulation tool for environmental modelling. Such simulation models can be used to 'close the loop' to provide a system under test with feedback of controlled system response.

Tool H A System Test harness, possibly configurable, to stimulate and record system responses.

Tool I A tool that compares and/or analyses results from a system simulation and from the actual system itself.

These generic tools are described below and some existing tools are identified.

Tool A Statemate supports the outline requirements of this tool. Statemate's modelling conventions are well suited to modelling the behaviour of control and protection systems. Of particular use is the ability to rapidly develop and test through execution, the logic associated with multi-channel systems required for high integrity applications.

Tool B System requirements related to reduction and removal of hazards must be demonstrated. In order to test the system functions that provide such protection, they must be identified as such. In decomposing a system into manageable units it is possible that a number of units combine to provide protection. In a large complex system there is a need to have some means of assistance in the creation and maintenance of such records. RTM is typical of such tools.

Tool C The test regime assessment model as proposed in Chapter 9 meets the requirement and would be simple to produce using a readily available integrated spreadsheet, database and word processing package.

Tools D and E The AdaTEST and McCabe Toolset packages are considered to offer appropriate support in certain areas of white box testing. However, they are limited to particular types of software system, those coded in 3GLs such as Ada, C, Pascal. They do not apply to systems developed using problem oriented languages, such as Ladder Logic, Grafcet or Function Blocks.

Tool F When a large number of test cases are to be applied to a system, it is useful to have a system that can manage the test cases as they provide valuable support in the scheduling and automatic execution of large numbers of tests. Such tools can be readily built using standard database packages.

Tool G This type of tool could also support the requirements of Tool A. Matrix-X is the typical of a tool well suited to modelling plant in terms of its physics, while Statemate is better suited to modelling control logic. It can be argued that each type of tool can support elements of the other's functions, each has its strengths in different areas. Integration of such tools is an area of interest as this would provide for control system and environment modelling using representations that are appropriate to each domain.

Tool H Two sets of test harnesses are required to stimulate and record system responses. The first being to test an application via its plant interfaces, the second being a total environment simulation for a piece of software, where the software is not executed on its target hardware. This serves to illustrate the fact that test harnesses will vary as testing moves from software testing to system testing. Clearly test harnesses have to be developed specifically for an application. This could lead to considerable extra expense for one-off and low volume systems. However, the development of system test harnesses capable of automatic execution are essential when random testing techniques are to be applied. Even if random testing techniques are not used it may be necessary to develop test harnesses to ensure that required timing of input signals is achieved.

Some organisations may use the same basic configuration of hardware for each system they produce, with differences being confined to the configuration of the I/O with the functional behaviour of the system, determined by the application logic. In this case, a 'standard' simulator could be defined for environmental simulation (depending on the simulation boundary chosen). Thus for each project, the complete simulator is not developed, but only the configuring software and the test definitions (which define the tests to be applied via the simulator) need to be changed. This approach could reduce the cost of using environmental simulation. This cost could be reduced even further where some scheme of automatic test case generation is developed. This would also provide the benefit of reducing the scope for introduction of errors into the test definition process.

Tool I When a large number of tests are to be performed, some means of automatically checking the predicted and actual results are desirable. It is considered that such tools can only be developed for specific applications and therefore it is unlikely that such tools will be commercially available.

For Tools C, F, H and I the requirements are regarded as being so specific that it is highly unlikely that any commercially available tools will ever exist. That is not to say that they have no use, rather it is to imply that the responsibility for their development lies with the organisations involved in the testing of systems. In such cases an appropriate development lifecycle must be adopted for these support tools and the development must be planned to coincide with the appropriate stages of the main system development. These development issues are highlighted in the discussion of the dual lifecycle model in Chapter 2.

A selection of commercially available tools that provide support for the requirements of Tools A, B, D, E and G are listed in Table 5-2 below.

Table 5-2 : Commercially Available Tools

Tool Category	Tool Names	Supplier
A	Statemate	i-Logix
	TeamworkSIM	Cadre Technologies Ltd
B	RTM	Marconi Systems Technology
	TeamworkRqt	Cadre Technologies Ltd
D & E	LDRA Testbed	Program Analysers Ltd
	McCabe Toolset	Instrumatic Ltd
	AdaTEST	IPL
G	Matrix-X	Integrated Systems Inc. Ltd

The commercially available tools identified here, a selection of which are discussed in more detail later, are a representative selection only. They have been included to illustrate the general problems and typical solutions found in safety-related system development.

5.4 Safety and Quality Requirements for Test Tools

5.4.1 Safety Requirements

Recent standards produced by the UK MoD, Interim Defence Standards 00-55 and 00-56, require that any tools used for automated verification and validation of safety-critical software shall themselves be subjected to proper validation and verification. Clauses 36.1 and 36.2 of the Interim Defence Standard 00-55 (part 1) are reproduced below:

Clause 36.1

> *'All tools and support software used in the development of the safety critical software (SCS) shall have sufficient safety integrity to ensure that they do not jeopardise the safety integrity of the SCS'*

Clause 36.2

> *'The safety integrity requirements of the tools and the support software shall be established by a hazard analysis and safety risk assessment implemented by the systematic application of Def Stan 00-56. This analysis shall define the safety integrity required of each tool in view of the role of the tool in the project. Criteria shall be defined for the assurance and evaluation of tools and support software for each level of safety integrity used in the project, either by direct definition or by reference to existing standards.'*

At present few, if any, of the commercial tools available meet the above requirements. These defence standards do not provide any guidelines on the conduct of hazard analysis on the tools that are bought as 'off the shelf' packages.

Clause 36.2 indicates that the safety integrity requirement for tools is dependent on the integrity of the system under development. It is perhaps reasonable to propose that a tool must have an integrity level which at worst is one lower than the level assigned to the product under development. Therefore, a tool developed to integrity level 2 could be used for developing software with integrity levels 1, 2 and 3, but should not be used for systems where safety analyses suggest a level 4 rating for the software under development. The rule of thumb given above for the integrity of the tools used is observed widely in UK industry. There is an informal consensus that 'one level lower than the product' is a prudent discipline.

5.4.2 Quality Requirements

The preceding parts of this chapter have discussed the functional requirements for test tools to support the development of a safety case. It is inevitable that part of the safety argument will build upon the quality arrangements to which the test tools have been developed or provided under. It is therefore instructive to summarise what advice and guidance the relevant quality standards give on test tools.

The following paragraphs present a review of the references to software development tool testing contained in a number of current and proposed standards:

DOD-STD-2167A, Para 4.3.2 Software Test Environment

> 'The contractor shall establish a software test environment to perform the FQT (Formal Qualification Testing) effort. The software test environment shall comply with the security requirements of the contract. The contractor shall document and implement plans for the installation, test, configuration control, and maintenance of each item of the environment. Following installation, each item of the environment shall be tested to demonstrate that the item performs its intended function.'

It has to be assumed from this standard that the test environment includes software tools. The standard groups the software test environment with the other testing requirements and again leaves the contractor to define the test plan.

ISO 9000-3, Para 5.7.2 Test Planning

> 'The supplier should establish and review the test plans, specifications and procedures before starting testing activities. Consideration should be given to...
>
> d) test environment, tools and test software.'

This standard indicates that the test environment and tools are different items. However, it does indicate there should be a test plan for tools. Once again there is no detail given.

RTCA DO-178B, Para 12.2 Tool Qualification

> 'Qualification of a tool is needed when processes of this document are eliminated, reduced or automated by the use of a software tool without verification of its output.'

This standard devotes an entire section to the testing and qualification of tools. The main points to be considered are: testing against its requirements, coverage analysis against requirements, and structural analysis appropriate for the tool's software level. The other quality issues relate to the planning, configuration management and review of the tools.

Interim DEF STAN 00-55 (Part 2)/1, Clause 36 Tool Support

> '36.5.1 Validation and evaluation of tools should be carried out by an approved third party using an internationally recognised test suite, acceptance criteria and test procedures, taking into account the operational experience of the tool. In cases where these services are not available (e.g. for clerical tools) the Design Authority should undertake, or

> *commission, an evaluation and validation and provide arguments that the tools meet a required integrity level.'*

This standard states that each tool used should be given an integrity level. The highest level of tool qualification would involve the tool being initially developed to the DEF STAN 00-55 standard. Other levels are to be qualified by the Design Authority. This then has two extremes:

- full use of formal methods in the design, development and verification of the tool
- qualification of the tool to show it meets the lower integrity level. No guidance is given for this.

The problem with existing tools is that they probably have not been designed, developed and verified using formal methods. Additionally, an internationally recognised test suite acceptance criteria and test procedures do not yet exist.

The tool types are classified into four classes and guidance is given as to which tools fit into which integrity level.

IEC 1508

> *'A certified tool is one which has been determined to be of a particular quality. The certification of a tool will generally be carried out by an independent, often national, body, against independently set criteria, typically national or international standard. Ideally, the tools used in all development phases (definition, design, coding, testing and validation) and those used in configuration management, should be subject to certification. To date only compilers (translators) are regularly subject to certification procedures; these are laid down by national certification bodies and they exercise compilers(translators) against international standards; such as those for Ada and Pascal.'*

This standard concentrates mainly on compilers and languages. It does not go into detail as to how tools are to be qualified. It really only states that to certify a compiler it should be able to pass a test suite, and thus does not mean the compiler is verified.

In general all of the above standards mention the qualification or verification of tools in one form or another. The most comprehensive standard describing the testing of tools is RTCA DO-178B. This standard is still dependent on much interpretation, e.g. 'What are the acceptance criteria for the qualification of tools that have been used for a number of years without a serious fault?'

Tool qualification levels need to be defined and recommended to suppliers. These levels depend on the type of tool and its purpose, e.g. the qualification level for interface simulation tools may be higher than system simulation tools. This is because the environment in the first case needs to be simulated as well as the inputs and outputs.

5.5 Statemate

5.5.1 Description

Statemate assists the development of clear, accurate, graphical specifications that are expressive and flexible but are based on a precise formal semantics [26], [27]. This type of rigorous specification is the foundation for reliable and predictable systems and carries through the entire development life cycle forming the basis for successful design, implementation and testing. The specification model can be executed and analysed to provide a clear understanding of how the system will behave even before the design phase begins – i.e. dynamic testing methods can be applied to seek faults before an erroneous specification is developed further.

Statemate is effective for reactive systems whose behaviour can be complex and their specification consequently difficult. Such systems exhibit distinctive characteristics:

- continuous interaction with the environment via asynchronous inputs and outputs which can change unpredictably at any time
- the ability to respond to interrupts (i.e. high priority events coming from the environment when the system is busy with another activity)
- their operation and reaction to inputs must often reflect stringent time requirements
- the use of different scenarios of operation dependent on current mode of operation, previous behaviour, etc.
- they are based on interacting processes that operate in parallel either actually or conceptually.

A Statemate model consists of three related and complementary hierarchical views of a system that together form a co-ordinated specification model that is comprehensive, clear and precise:

Functional activity charts which form a functional decomposition of the system – e.g. as for conventional data flow diagrams;

Behavioural state charts representing dynamic control over time in the system environment that permit the capture of concurrent behaviour. They extend normal state diagrams to express hierarchy, concurrency, disjoint states and broadcasting of transition conditions between states;

Physical module charts that are used to assign implementation responsibility to the hardware and software components for the system's functional and control aspects.

Checking functions are provided within Statemate to test for violations in syntax or semantics that examine the correctness (checks for inconsistencies) and completeness (checks for redundancy or deficiencies) of the different system views and their interrelationships.

The Statemate Analyser is used to execute the system model and permits:

- interactive simulation where the user plays the role of the environment, generating events and setting conditions and data values, to dynamically test the operation of the system
- batch simulation of scenarios that represent typical desired behaviour of the system
- dynamic tests which determine whether specific undesirable behaviour could occur with the system as specified via exhaustive testing. The range of tests includes dynamic reachability, non-determinism, deadlock conditions and usage of transitions.

Also provided is the ability to graphically create soft panels to allow the user the opportunity to validate the behaviour of the specification through a friendly and familiar interface of sliders, knobs, buttons, meters, etc.

Statemate incorporates its own configuration management system and access control system to support controlled development.

5.5.2 Statemate Benefits

Early Executable Specification Permits Dynamic Test Methods to be Adopted
Statemate, through its ability to execute the specification, allows methods that were appropriate to dynamic testing of software to be brought forward in the development lifecycle. Black box test techniques can be applied to a Statemate Model and, to a lesser extent, white box techniques can also be applied. Such testing will clearly not be related to any structural coverage of final executable code, however, these techniques can be used to provide evidence that some level of systematic testing of the specification has been completed. It is argued that introduction of 'active' testing much earlier in the development lifecycle will promote the detection of errors in the specification.

Easy to Simulate Multi-Channel Systems and their Interaction Statemate is very effective for modelling multi-channel systems that are required to interact. The behaviour of each channel may be complex, and the difficulty in analysis of behaviour is compounded when the combined system's behaviour is considered. In such circumstances an executable model of each channel can aid the analysis considerably. Statemate supports the development of generic specifications that may then be instantiated. Thus a generic model for one channel may be constructed which can then be used for each subsequent channel required to complete the model. Only the actual parameters passing between channels need to be added to complete the model. The development of generic models offers the opportunity for reuse of specifications.

Graphical Representation of Behaviour Statemate's graphical notation makes the specification of complex behaviour concise. It provides an implementation independent notation that promotes communication between members of a multi-disciplinary team, for both review and development activities. Additionally Statemate's graphical panel interfaces allow the developer to provide 'customer friendly' interfaces allowing the customer to actively test a prototype early in development.

5.5.3 Statemate Drawbacks

Design Phase Support   Support within the tool for the development of design specifications is limited. Module charts can be adopted to specify a design architecture. However, model checking can only be used effectively if the design structure has been specified before the functional behaviour. In some circumstances this may be an appropriate way of working. If a new model is constructed specifically for the design stages there is no support to maintain traceability between models.

Support for Random Testing   Execution time and data conversion overheads preclude the use of Statemate as an 'oracle' for random testing. This is a common weakness of all general purpose simulation tools and is not specific to Statemate.

5.5.4 Review Conclusions

Statemate is considered to be a useful tool for the development of functional specifications.

A test schedule should be developed for a Statemate model based on black box and white box techniques for software system testing and that evidence of completion of such testing is recorded.

Usage guidelines are considered necessary to help users construct models. The target hardware and the implementation language may make it necessary to impose certain constraints on either the constructs used or the way in which logic is specified using Statemate. Imposing these constraints will lead to better traceability between specification and implementation. This improved traceability will aid the testing process. Because the translation process between analysis and design phases in not well supported by CASE tools the imposition of these constraints to improve traceability become more important.

Statemate is considered less useful for the development of design specifications. While it is possible to do so, the success of such an activity is very much dependent on the user's ability to reproduce the functional behaviour in a form that is appropriate to the implementation languages and/or architecture. It is suggested that a more appropriate way of working would be to adopt a suitable design representation, for which other tools may exist, and to maintain traceability links between the functional and design specification representations.

5.6 Requirements and Traceability Management (RTM)

5.6.1 Description

Requirements and Traceability Management (RTM) enables requirements to be engineered, tracked and traced throughout the entire project lifecycle – from requirements definition to system delivery and maintenance, regardless of the particular development process in use. It does this by allowing the specification of the process in a graphical manner, by defining the relationships between classes of information produced in the lifecycle in a class definition diagram. In essence, any traceability matrix can be modelled and maintained within RTM. Classes represent individual classes of data. Each of these classes can have user-defined attributes to hold specific information about the item the class represents. Once these classes have been defined, then instances (objects) of these classes can be created by the user and entered into the tool, e.g. individual requirement statements, design specifications statements or test cases references. Separate editors are provided to handle two generic types of class instances, requirements, and all other types of data.

Requirements may be grouped according to user-definable keywords. These groupings can be arbitrary and reflect the user's view of the system's functionality. They can be used for partitioning and assigning design tasks to differing teams or individuals.

RTM consists of a series of software options or modules that can be purchased individually according to the needs of an application. In addition to the core modules included in every system, there are options for graphical requirements management (for storing graphical and/or tabular requirements extracted from an Interleaf requirement source document), database partitioning (for distributing the RTM database across several machines and/or sites enabling physically separated teams to work effectively on the same database), an interface to the configuration management tool PCMS and interfaces to the CASE tools 'Software Through Pictures' and 'Teamwork'.

Once the project data has been entered, it can be extracted from the RTM database according to a set of rules defined in a script. RTM scripts can also contain formatting instructions suitable for use with Desk Top Publishing systems such as FrameMaker or Interleaf. Rich Text Format (RTF) output suitable for some word processors is also supported, as are plain ASCII reports. Scripts are generated using a script generation tool. The actual script generated is very similar to SQL but its generation is simplified by the use of graphical tools using the so-called 'query by example' paradigm. The scripts are run using a documentation tool. This can run scripts that have just been generated by the script tool, or scripts previously saved to file. Previously written scripts can also be run from an operating system command line without invoking the RTM tools, by using a utility provided for the purpose.

The RTM application uses Oracle version 7 as its underlying database – a run-time version of the database management system is supplied. A tool called the Database Management Utility (DBMU) is used to administer the project database. The project database is central to the RTM environment, containing the project data captured and manipulated throughout the RTM lifecycle. The DBMU provides a graphical interface that is intended to simplify database management.

5.6.2 RTM Benefits

Provides an Environment for Requirements Engineering RTM provides a framework for undertaking requirements engineering and enables requirements to be edited, focused (combining requirements), and expanded (split into subrequirements) reasonably easily, whilst maintaining traceability back to the original requirements as either entered by the user, or extracted from a source document. In particular, its strengths lie in managing large, complex projects whose requirements need significant engineering, possibly by several teams. It has the ability to act as a central data repository and hence aid project handover from function to function (or indeed company to company).

Allows Design Items to be Traced to Original Requirements The tool is designed to be flexible and not impose any particular design process or methodology on the user. The user can define any traceability matrix by describing graphically the relationships between any classes of information in the development process. These may include requirements, design specifications, test specifications, review notes, etc.

Powerful User-Configurable Reporting Facilities Some simple reports are predefined and can be produced by executing tool menu options. Other reports can be generated using a utility to produce SQL-like database queries. These can be arbitrarily complex to produce the required report. Once created, these report-producing queries can be saved and run at any time in a variety of forms.

5.6.3 RTM Drawbacks

Potential Project Overheads There is a potentially unacceptable overhead in using the tool on relatively small projects whose requirements are already well understood at project inception and which have small development teams. In this case, setting up the project, including the definition of all the classes and relationships, might possibly be too much of an overhead with little potential gain. A set of standard working practices and procedures would help to address this, but there seems to be no in-built way of defining a standard template for a company's development process as defined by a class definition diagram, so each new project would have to be set up individually. It might be possible to achieve this by operating at the database level and saving/copying/restoring databases – this is not the best of solutions.

Ease of Use The tool is complex. Generally, the advantage of flexibility can also act as a disadvantage by increasing the functionality (and hence potential complexity) of a tool. In the case of RTM this could be reduced by improving its user interface. As an example, there seems to be no easy way of performing database backups or baselining project databases – this must all be executed from the Oracle database system command line, even though a database management utility has been provided, which at first sight could have had these facilities added.

This problem may be a drawback in the use of many tools. This was seen in the use of RTM and would impact on several areas in the following ways:

- contributing to data entry errors, affecting quality (and potentially safety) of the final product
- adversely affecting the perception of the benefits of the use of such a tool and hence the willingness to use it
- increasing the costs of the day-to-day use of the tool
- increasing training costs.

Useability should be a prime requirement placed on any tool or set of tools. Improving presentation, control and data aspects will all positively affect not only the use of an IPSE, but also that of stand-alone tools which are composed of a set of interworking modules using a common database. RTM illustrates this problem, where more thought on useability would improve a complex tool that executes its basic function well.

Limited Integration with Other Tools   Integration is limited to two structured analysis and design tools and a configuration management tool. Users of other tools must provide their own work-arounds for this problem and enshrine them in working practices and procedures, probably resulting in a suboptimal solution. The alternative is to ask the tools' vendors to provide special software solutions. This is likely to be expensive.

5.6.4 Review Conclusions

RTM is a complex tool and the effort required for its introduction and training of staff needs to be carefully balanced against its advantages.

A set of working practices and procedures requires to be established to describe how users should apply the tool in their particular circumstances. These should ideally be based on experience gained from using the tool on a trial project.

RTM supports traceability throughout the development lifecycle. However it was found difficult to use primarily due to its configurability. For modestly sized projects, typically run by one company rather than a consortium, the use of such a tool may not be appropriate.

5.7 AdaTEST

5.7.1 Description

AdaTEST Harness (ATH) for Dynamic Testing The AdaTEST consists of packages of directives, which enable a developer to construct a *test script* in the form of a main procedure; this main procedure uses the functions provided by the package under test. The *test script* contains *test cases* and these *test cases* have two vital components:

- initial conditions and
- expected results.

Figure 5-1 shows a typical test environment.

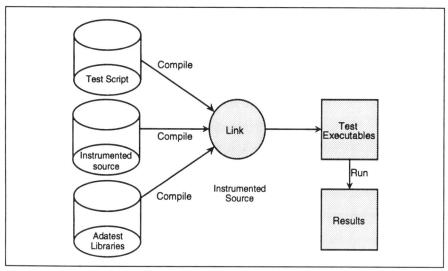

Figure 5-1: Building and Running a Test

The test script that constitutes a *Ada Test Harness* (ATH) is constructed from *generic* packages and directives. The *generic* packages are instantiated with the *user-defined* data types. The test script is compiled and an executable image is built by linking the test script, the package under test and the ATH library. The software under test is executed when the test harness calls the procedures and functions that are exported by the package; the effects of the calls on the software environment are checked. When an ATH is executed, a result files is produced that contains full details of any changes in the state variables and other data structures in every step of the test; any diversion from the expected values is recognised as a failure and highlighted in the result file. A table summarising the results from the entire test including an overall statement of *pass* or *fail* is written in a separate file. This file can be viewed using an ordinary text editor; the results can be exported to a spreadsheet package to produce charts, bar graphs, and other visual effects.

The module under test can be analysed using the AdaTEST Analysis packages (ATA), which comprises:

- an instrumenter program
- an additional package of test directives.

Figure 5-2 shows the usage of these facilities.

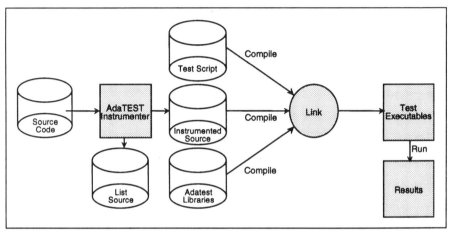

Figure 5-2: Use of ATA

The source module is processed with the AdaTEST instrumenter, which generates a list file containing:

- an annotated version of the source file containing numbered statements, decisions and exceptions and a static analysis repor
- an instrumented source file, which is a modified source file containing additional software probes so that execution progress can be monitored by AdaTEST is produced.

AdaTEST Analysis (ATA) for Test Coverage and Static Analysis This tool consists of a set of additional library directives and source code analysers. Using these, the developer can analyse a source code file, rebuild this test executable, and by adding appropriate directives to the test script, obtain coverage and complexity information in a number of different formats. It supports the complexity measuring metrics such as the Halstead [28] and the McCabe [29] complexity measures.

Using the AdaTEST Analysis tool three types of coverage statistics can be obtained. They are:

- *decision* coverage, which is a measure of the proportion of decision outcomes executed

- *statement* coverage, which is a measure of the proportion of statements executed during the test run
- *exception* coverage, which is a measure of the proportion of exception handlers executed.

AdaTEST Analysis gives access to three metrics, they are:
- McCabe's measure and Myre's extension
- Hansen's software complexity by the pair
- Halstead's software science.

Environment Simulation During Dynamic Testing AdaTEST simulates the environment of a module under test, with stubs and drivers being used to simulate the environment of a module. It does not provide any packages or libraries for constructing large scale environment simulator for systems or subsystems' integration testing.

5.7.2 AdaTEST Benefits

Ease of Use Ada test harnesses (ATH) are written in a test script language that is very comprehensible; the syntax and the semantics of this language are very similar to Ada. An average programmer with good working knowledge of Ada should be able to master this tool.

Representation of Test Data Test data is embedded into the test script. A template for the test data of individual test should be provided (a separate file for each test). The test result is recorded in a separate file (matrix.atr). These include the percentage of code executed by the tests, results of the static analysis and complexity of the module.

The test results can be used as proof for verification, the result files are generated in ASCII text format.

Validation and Verification of AdaTEST The tool vendor, IPL, indicates that this tool has been developed within IPL's ISO 9001/TickIT approved quality management system.

Platforms This tool is available on a large number of platforms, they include Appollo(Unix), HP 9000 series, IBM PC(MS-DOS, OS/2), PS/2, SUN(Unix), VAX(VMS), IBM RS6000(AIX), IBM SYSTEM/88(VOS).

5.7.3 Ada TEST Drawbacks

Testing Double Insulated Packages

Testing double insulated [30] packages is difficult; AdaTEST does not allow the tester to check data that are declared in a double insulated package. *Double insulation* is highly recommended for packages that are safety-critical, for instance packages that control hazardous input/output. For example if a package body PACKAGE_A contains another package PACKAGE_B (both specification and body) then data declared within package PACKAGE_B is not visible to other packages that are using PACKAGE_A. If an Ada package handles safety-related devices then the handler for each input/output device is enveloped into a package containing the safety interlocks, and the package specifications are only made available to those operations that include the necessary safety checks. For example if package PACKAGE_B handles safety-related input and output then if only package PACKAGE_A performs the safety checks then the PACKAGE_B specification and the body is declared within the body of PACKAGE_B.

The AdaTEST vendor, IPL suggests two methods for testing PACKAGE_A:

- the package PACKAGE_B can be compiled as a separate library and this library can be called from PACKAGE_A. If PACKAGE_B is made a separate library then its data is available to all external packages; it can be tested standalone using AdaTEST; or
- extra stubs can be added in the specification of PACKAGE_A that can access PACKAGE_B's data directly. These extra stubs will be redundant once the system tested fully. Figure 5-3 shows one of these possible methods to test package A and B.

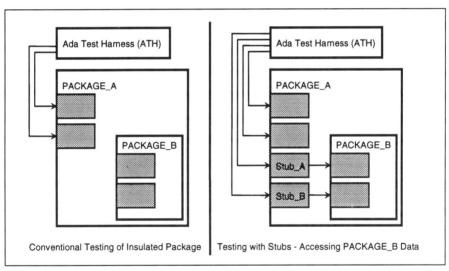

Figure 5-3: Testing Facilities of a Package Declared within Another Package

There are two disadvantages to these approaches:

- A *double insulated* package encapsulates critical data and prevents misusage by the external packages. Separate compilation of PACKAGE_B will expose its encapsulated data to the outside world.
- If additional stubs are used then extra redundant code will occupy extra memory space and other resources. For embedded systems where

memory is limited this would not be very desirable, also the package will not be maintainable.

User Interface The tool does not provide a graphical user interface – which can enhance testing. However, it is adequate for testing modules at low levels.

When AdaTEST was used to test packages which contain Ada tasks that have interaction with the tester it was found that the output from the tool interferes with the output from the modules under test. For this kind of module a real-time graphical user interface is highly desirable.

Effect on Total Productivity of Development Lifecycle The test harness for the modules can be produced from a template but setting up the tests is very time consuming. The productivity (in this case number of modules tested per day) will not increase, and the test harnesses and stubs must be hand coded which will take a considerable amount of time. It does not provide any facility to automate this process. The cost of using AdaTEST to demonstrate Integrity needs to be justified.

Support for Real-Time Testing AdaTEST provides limited capabilities for testing modules in real-time on a host system and real-time behaviour of a module under test is implementation dependent. The tool was evaluated on VAX/VMS system. The granularity of time is dictated by the run time system (Ada run time system) and the smallest measurable time granule is ten milliseconds. This is adequate for some *soft* real-time applications but not suitable for *hard* real-time systems which have operating cycles less than a hundred milliseconds.

A *soft* real-time application demands system response within a finite time but failure to meet this demand will not cause catastrophic accidents. A *hard* real-time application demand responses from the operating system and the environment within a finite time and failure in meeting deadlines may cause hazardous situations.

5.7.4 Review Conclusions

AdaTEST tests instrumented source codes and the environment of the software affects its behaviour. The behaviour of the software is dependent on a number of factors including – the hardware platform, the operating system, execution environment; the instrumented source codes have different execution environment than that of the target system. The software should be tested in an environment that has affinity with the target system.

AdaTEST is widely used in the safety-related software development community and is available on a wide range of platforms. It carries out a range of static and dynamic test activities well.

AdaTEST is easy to learn due to the similarity of its test script language to Ada. This provides a significant benefit to the user.

5.8 Integrated Tool Support

The main objective of an integrated environment is to increase productivity. Organisations distinguish between 'horizontal' tools, which cover many phases of the lifecycle and 'vertical' tools which are applicable in only one or two phases, as illustrated in Figure 5-4.

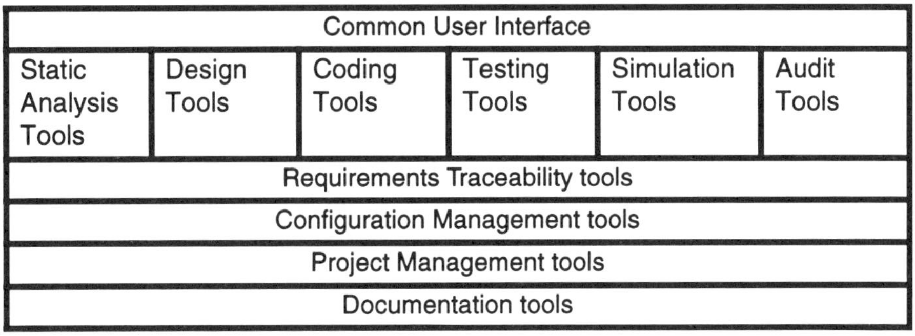

Common User Interface						
Static Analysis Tools	Design Tools	Coding Tools	Testing Tools	Simulation Tools	Audit Tools	
Requirements Traceability tools						
Configuration Management tools						
Project Management tools						
Documentation tools						

Figure 5-4 : Conceptual View of Classes of Tools

There is a consensus amongst the systems and equipment companies, that integrated project support environments (IPSE) do not live up to their potential. The main problem with existing IPSEs is that 'integration' is ill defined. An IPSE is not just a collection of tools and an operating system.

Careful investigation of the capabilities of each potential tool should be carried out before decisions are made on integration of the tools. For example, Statemate has an additional utility which supports requirements tracing within the tool. Therefore it is questionable whether integration between tools should be attempted if an adequate facility is already available by other means.

5.9 Tool Selection Criteria

Previous discussion in this chapter has been general and not project specific. Clearly it is important that tools are selected on the basis of their value to the project being considered.

Purchasers of tools need to consider very carefully how well tools fit their process. If the fit is poor, they may need to decide whether they are willing and able to change that process to fit the tools. In the field of safety-related systems, the development process forms part of the safety justification of a system. Until there is a universally accepted development process one must accept that the tool vendors are unlikely to offer tools configured to be completely compatible with a particular development process. The approach that tool vendors take currently is to offer configurable tools. The configuration work can be very complex and add considerable expense to the project so it may be expedient to accept that the project may have to adjust the development process to fit a non-configurable tool rather than to adopt a configurable tool. The costs of such changes should be considered before tools are adopted.

Establishing Selection Criteria

The selection criteria can be divided into four groups:
- general
- environment-dependent
- tool-dependent functional
- tool-dependent non-functional.

General Criteria The main objectives of incorporating new test tools into a project are to increase development productivity and software quality. The quality of the software under test depends on the quality and reliability of the tools used for testing. Most companies selling tools in the market have incorporated quality assurance procedures within their organisation that are compliant with the ISO 9000/TickIT or BS5750 standards. The user community also suggests that although a company may have been accredited by standards organisations this does not necessarily mean that the tools they produce will meet the purchaser's quality requirements. At present the metrics that are used do not give adequate measures of productivity and quality gains. The purchaser's quality plan is influenced by the standards they follow to develop their product. The companies developing safety-related systems come from a number of industrial sectors, each with its own standards.

Environment-Dependent Criteria This include issues such as the cost of the tool and cost/benefit ratio. Certain tools may provide an integrated development environment but the cost of the tool may outweigh the benefit.

Organisational behaviour also affects the usefulness of a tool. Studies have shown implementation can fail when companies try to incorporate a new method or tool without allocating time and resources for training. Most tools provide graphical user interfaces and training often amounts to a quick walk through of the menu system that can be accomplished on the job, but learning to use a tool efficiently and to exploit its features requires classroom time.

When a new tool is incorporated it may affects the existing tool interconnections because it uses different input and output formats.

Tool-Dependent Functional Criteria Project managers create a 'wish list' that contains desirable functions for tools but getting all of these functions may not be practical. It is much easier to select appropriate tools by concentrating on a few, high priority functions. The functionality of the tools should be selected from the existing tools but they should arise from the genuine needs of the tester and managers.

Tool-Dependent Non-Functional Criteria These are the measurable properties, attributes of the tools. For instance the performance of a tool is often measured in response time, other attributes may include ease of use. Terms such as *user friendly* are ambiguous; most selection criteria for test tools include a requirement for a user friendly human computer interface. A friendly user interface does not necessarily mean increased productivity. There are many different kinds of users, for example, frequent, casual, and professional. Most tools tend to have graphical user interfaces that are very intuitive. These GUI based tools operate in a very complex environment that can effect the way human beings interact with them. Research has shown that the most productive tools are those which provide a very simple HCI interface and perform the most undemanding jobs, i.e. where there is very little intellectual input from the operator.

5.10 Guidance

- Assess the requirement for test tools for the project using the toolset equivalents detailed in 5.3.

- Determine the safety integrity level and quality requirements for the tools and assess the tools against these requirements.

- Consider the use of a formal set of tool selection criteria. Outline suggestions are given in 5.9.

- Test tool integration should be viewed with caution. Co-operating tools that aid the human tester are preferred.

- A project specific investigation into the suitability of tools is essential when an organisation is considering the use of tools to assist in the testing of safety-related system.

Chapter 6
The Use of Simulators

This chapter discusses the use of simulation as a test method for programmable electronic systems (PES) and considers how it may be used to assist safety justifications. Two models of software environment are described, namely the set theoretical model and the stochastic process model.

When validating safety-related systems using environment simulators any inference made regarding safety must be conditional upon the accuracy of the simulation. The accuracy required of the environment simulation for systems of different integrity levels and complexity, and the factors that determine whether environment simulation is an essential component of the validation process are discussed.

Evaluation of simulation accuracy is considered as the approximation match of the safety-related system and its environment. Two methods of approximate matching are described, the equivalence relation approach and the metric space approach. The concepts and terminology of these approaches are introduced.

6.1 Introduction

In many programmable electronic system development lifecycles, simulations are used for studying system behaviour or to test the system in an artificial environment that provides a limited representation of the actual environment. A simulator is an abstract model of a system and can be developed using various techniques.

Environment simulation involves modelling and different modelling formalisms are applied for different environments. In this chapter a brief description of different types of formalism available will be given. Software based environment simulation plays an important role in the development of safety-related systems, particularly in the validation and verification phases. In this chapter descriptions of two models of software environment, the set theoretical model and the stochastic process model, are provided.

Any model based system can be considered to have two major components: a behavioural model and an environmental model. The former describes the expected behaviour of the system; it models the system behaviour that is described in the functional specification of the system, while the latter describes the interaction of the system with its environment. A number of types of environment simulators are described in this chapter, and recommendations of their usage in the system development lifecycle are given.

A simulator is a model that imitates the functions of a man-made system or a natural process (e.g. the weather system). In a simulation study only those functions that are of interest to the developer are modelled by the simulator. When validating safety-related systems using environment simulators any inference made regarding safety must be based on the accuracy of the simulation result. It is therefore important to assess how accurately the environment simulator mimics the actual environment. The evaluation and assessment of the simulation accuracy can be considered as the approximate match of the safety-related system and its environment. Two methods of approximate matching problems are described:

1   the equivalence relation approach
2   the metric space approach.

6.2   Types of Environment Simulators

A safety-related system must be developed using rigorous procedures and standards. Most programmable electronic systems have a development lifecycle similar to the 'V' model. The relevant draft IEC standard [6] recommends that a safety-related system should have an overall safety lifecycle. It suggests that verification activities should be performed in each phase of the lifecycle before the next phase can commence. The development lifecycle of such a system is a part of this safety lifecycle. Simulation and emulation may be carried out in various phases of the development lifecycle as a means of verification and validation of the processes namely:

- the system requirement analysis phase
- the design and construction phase
- the test and integration phase.

Simulation also has an important role to play during the post-implementation phase where changes in plant requirements or problems with plant operation may be encountered. The impact of any modifications due to a change in requirements or operational problems must be identified and understood before any recommendation for the change is made. Simulation is the most effective way of achieving this as it avoids excessive use of the plant in testing and the associated risks to plant operation.

In each of the phases described above, different types of simulators and emulators are required to aid testing and verification. The following classes of simulators, which support testing at various phases during the whole development lifecycle, can be identified:

Specification Animators

Before the system is implemented it must be defined as a set of specifications that are checked by a review process or by conversion to an executable form. Various conditions can be presented to the executable specification to elicit responses that can be confirmed by review as being correct. This conversion of the specification to another form may also be useful in identifying ambiguities. Specification animators are usually used in the early stages of the development lifecycle; they are applicable in the hazard analysis and requirement analysis and architectural design phases. For safety-critical systems, animation of the software specification during the requirement analysis phase is highly recommended.

Component Environment Simulators

Once components have been developed and implemented they can be tested. The role of a simulator in this phase of development is to provide stimuli to the components under test, exercising them to the degree required to confirm that each component meets its specification. Component environment simulators exemplified by a test harness for a low level software module, are normally used during the module construction and testing phases.

Target System Drivers

Once all the components that comprise a system have been tested they may be integrated into the target system. There are a number of concerns regarding the simulation of the plant and the simulation of the system under test. Simulators may be classified as follows:

Simple static signal injection This type of simulation aims merely to exercise the data acquisition aspects of the system, to establish that the transfer from electrical signal to computer system memory location takes place. It can be effected by switches and voltage or current sources. This type of simulator is usually used during the module integration and testing phases.

Simple dynamic signal injection In many real-time control system designs there may be a requirement to test the response time of the data acquisition subsystem. For such tests the inputs can sometimes be adequately generated using standard test equipment. These types of simulators can be used for standalone subsystem testing on a target computer.

Plant behaviour reproduction (step mode) The levels of simulation described above are intended only to test the low level functions of a PES. The next stage is to attempt to exercise the logic of the system in a single step mode. The simulator may well make use of the equipment described above, but the intent is quite different; it is to trigger system state transitions. Some knowledge of the system logic and the plant behaviour is required in specifying the signals and the sequence in which they are to be injected. This type of simulator can be used for system integration testing, where the subsystems are connected to build a complete system.

Plant behaviour reproduction (automatic mode) The aim of this simulator is essentially the same as that of the step mode, to trigger state transitions by injecting discrete signals, but in this level of simulation timing aspects may be introduced. At this stage the equipment used to generate the signals may be quite different because to meet timing and repeatability constraints the signal injection will be controlled automatically. This type of simulator can be used for open loop testing.

Plant behaviour reproduction (closed loop) The final level of simulator considered here is one in which the continuous dynamic behaviour of the plant in response to signals generated by the system are fed back into the system. This level of simulation would involve both continuous and discrete signals. The behaviour of the plant may be reproduced using a mathematical model that can execute in real-time, an analogue model, or possibly even a pilot plant in which the hazards are reduced or eliminated. This type of simulation is used during system validation. It also has a major role to play after system installation where it can be used to assess and validate changes to system requirement and to investigate system operational problems

The discussions above indicate that a range of simulators may be required during the development of a practical system. It is important, therefore, that a design approach is adopted which ensures that all simulation toolsets are developed, operated and applied in a consistent and co-ordinated manner throughout the entire lifecycle of the system.

A practical example of an integrated approach to environment simulation is the Workstation Based Engineering Simulator (WES) developed by Scottish Nuclear. This is a networked environment providing the basis of an integrated platform for environment simulation. It was originally developed for investigating, designing and dynamic testing of nuclear plant control systems. The WES can use in-house and commercial modelling tools for developing environmental models and offers the potential for using simulation at various stages during the development lifecycle.

An investigation of the potential of WES as a through-life environment simulation platform for developing safety-related systems can be found in [31].

6.3 Use of Software Environment Simulation in Testing Safety-Related Systems

Safety-related software must be thoroughly tested by test cases that accurately represent the real operational plant or system. For the following two reasons, it is often difficult to obtain adequate realistic test cases that represent the real operation of the software. Firstly, safety-related software is often developed in parallel with the design and implementation of its environment, the system in which the software finally executes. Therefore, it may be impossible to obtain test cases by recording the real operation of the system. Secondly, since the malfunctioning of such software could lead to unacceptable risk to human life and property, the software must achieve the required reliability the first time it is put into operation, consequently there are few chances to test the software in its real operational environment without threat of danger.

Software testing by environment simulation is an approach designed to overcome such difficulties. The main advantage is that the software can be tested by highly realistic test cases, but without threat to human life and property. Another important feature of the approach is that the software can be tested not only in the normal operation of the environment but also in adverse conditions representing the faulty operation of the environment. This allows the robustness and tolerance of the software to unexpected system behaviour to be assessed, as such testing could be too hazardous to be carried out in the real environment. Moreover, it could not be performed adequately in the real environment because the design of the system should have minimised the probability of the occurrence of hazardous situations. For example, in nuclear installations it is normal for systems to be designed to protect the plant from very low probability fault situations, that indeed may never occur during the lifetime of the system. However, for safety-related software such as that used for protection, testing in adverse plant conditions is essential. In these cases, simulation of the possible environment is necessary if realistic testing is to be performed.

The development of a software environment simulator starts with modelling the environment system. This model is then refined and finally expressed in the form of executable computer program.

In the following paragraphs two generic models of systems of software executing within an environment are discussed. The first generic model is a set theoretic model to capture the functional properties of software behaviour. The second is a stochastic process model to capture the probabilistic properties of software environment.

6.3.1 Set Theoretical Model

Figure 6-1 illustrates a model of a man-made environment, i.e. the environment of the subsystem C consists of other man-made subsystems (A, B, D and E). The fundamental problem to be solved by the model is: when the environment is switched to a simulator to test the software in subsystem C, how do we determine how representative such testing is compared with testing in the real environment?

Suppose that there is an observer who monitors and records the behaviour of the system, but the observer cannot interfere with the performance of the system.

The observer has a collection of observable events of the system in which he is interested. When any of these events happens, the observer records the event in detail. Only these events need to be taken into account to decide

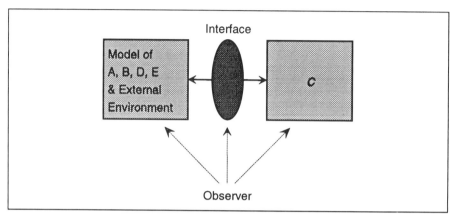

Figure 6-1: Monitoring and Recording the Behaviour of a System

whether or not the two environments are the same. Generally, there are three types of atomic events that is, events that cannot be divided into component events:

- the communication between the software and the environment, i.e. the passage of information across the interface. The set of communication events can be further divided into two subsets according to the direction of the passage of the information, which can be from the software to the environment, or from the environment to the software
- events that happened in the environment without participation of the software, e.g. the ringing of an alarm bell
- events that involve only the software, such as the program entering a particular state.

It is also possible that two or more atomic events happen at the same time, a compound event. In particular, the empty set is also considered as an event which means that no events take place at the moment. This event will be called silence and written τ. (The name and symbol of the empty event is borrowed from theories of process algebra.) Another special event is termination, written $\sqrt{}$, which means that the performance of the system terminates at the moment. If termination $\sqrt{}$ happens at a moment t, the only event which can happen afterwards is τ. The collection of events will be called the event space. In the remainder of this chapter, we assume that the event space Σ always contains these two special events.

However, it may not be sufficient to make a decision about the equivalence of two environments based on a sequence or sequences of events. As a very careful monitor, the observer also records the time when an event happened, but, there is a problem as to what precision time should be recorded. The choice of the precision depends on the system. However, given a precision, we have a time index set which contains all the time 'moments' at which events happen. Since software systems are artificial objects, there is always a moment that the system starts its performance. It is assumed that the observer always begins to record the behaviour of the system from the start moment and that the behaviour of the system is dependent not on the absolute time, but on time relative to the start moment.

It is assumed that the time for which the system can last is not bounded by any given limit and that the same time recording precision is used in the whole record of the behaviour of the system. It is also assumed that there will be a fixed sampling rate called *time resolution*, with each sampling having a time stamp. A time index set, characterised by two real numbers, the start time t_0 and the time resolution ρ, is created. Without loss of generality and for the sake of simplicity, it is assumed that every event takes no time to complete. Any *event*, which lasts for finite time can be split into two events, one representing its start the other representing its end. Its duration is simply the time period between the time stamps of the start and end events.

A particular operation of software within its (real or simulated) environment can be considered as a sequence of events indexed by the time index set. An instance of an environment of a given software is defined to be a function whose domain is the time index set and range is the set of events, these events in the set occur at times specified in the time index set. A software environment is, then, the set of all possible instances of this function.

It was stated earlier that a time index set can be characterised by the start time and the time resolution. When the time resolution is 0, the time index set is continuous, but when the time resolution is greater than 0, the time index set is a discrete set. According to the time index set, an environment can be regarded as discrete or continuous. Most software environments are discrete and digital. The instances of an environment still need to be decomposed into a sequence of segments to allow the study of the behavioural structures of software environments.

A segment of an instance of software environment is defined to be a function from a time interval to the event space. In another words the segment is a function whose domain consists of a slice of the time index set with a start and end time, and the range of this function contains a set of events. Each member of the time index set is related to an event by the function.

Since the time space is uniform, a segment defined on an interval can also be defined on any interval of the same length. Segments can also be concatenated together to form larger segments.

6.3.2 Stochastic Process Model

Some basic notions from the mathematics of stochastic processes are discussed below. For any given time index set a stochastic process is a collection of random variables. For each time stamp in the time index set there is a probability of the occurrence of any specific event and this probability can be expressed as a random variable ranging over the event space. When the time index set of a stochastic process is the set of natural numbers, the stochastic process is a sequence of random variables. A sequence of particular numerical values assigned to the random variables by appropriate use of a random device is called a realisation of the process or sample path.

A software environment described by a stochastic process over an event space consists of all the realisations of the stochastic process. However, not all the realisations have the same probability of appearing in practical use of the software. A realisation of a stochastic process is an instance of software environment in terms of our set theoretical model of software environments. The difference between the two models is that in set theoretical model, an 'instance' either belongs to an environment, or not. Whilst in stochastic model, an 'instance' is associated with a probability of occurrence in practical use of the environment.

Only discrete environments will be considered, i.e. the time index set is discrete, and without loss of generality, it is assumed that the time index set is a set of natural numbers.

In the simplest case, the random variables of a stochastic process are independent one of another. The states of such a system in the future is independent of either the past states or the present state. For most systems that arise in the practice, however, the past and present states have influence to the future states even if they do not uniquely determine them. Markov chains are such stochastic processes. In a Markov chain, the state of the future is influenced by the current state of the system but not the states of the past. Simulation of probabilistic software environments using Markov processes is described in [32].

6.4 Environment Simulation Accuracy and its Assessment Based on the Set Theory Model

The following paragraphs contain a discussion of the accuracy required of the environment simulation for systems of different integrity levels and complexity. The factors that determine whether environment simulation is an essential component of the validation process are discussed. Any simulation is only as good as the mathematical or physical model upon which it is based. It is important, therefore, to ask how *good* a model needs to be, the parameters that need to be modelled and the accuracy of their simulation. At present very little research has been carried out in determination of accuracy of simulators. The present approach to assessing the simulation accuracy is based on the approximate matching of the two environments. There are two basic approaches to the approximate matching problem, the equivalence relation approach and the metric space approach; a discussion of these approaches is given below.

Factors such as the intended use of the simulation data, the criticality level, the time scale, the resource availability and the expected benefit/cost ratio need to be considered to determine the appropriate simulation accuracy.

In general, the accuracy of a simulation is dependent on:

- The parameters that are simulated. The environment simulator must provide all input variables to the system (software) that can significantly affect its state. The range of the parameters must be carefully selected so that only the parameters values within specific range (oper-

ating envelope of the system) are used to simulate the environment of the system.
- The numerical accuracy of these parameters. The numerical accuracy is solely dependent on the instrumentation techniques used to measure external variables of the system and the computer system used to control the system.
- If the simulation is performed in real time then the speed of responses to the stimuli from the system is also an important factor in determining simulation accuracy. The real-time behaviour of a system can be recognised as either *hard* or *soft* real-time. A discussion of timing issues is given in Chapter 4. The granularity of the real-time is restricted by the processor clock speed and the operating system's capability of providing time to user tasks. At present most real-time operating systems can handle a time granularity of the order of milliseconds.

At present there are no metrics available that can be used to measure quantitatively how accurately the environment of a system is simulated or the effectiveness of the test data in finding errors in the system requirements. However, there are two basic approaches to the approximate matching problem, the equivalence relation approach and the metric space approach.

6.4.1 Equivalence Relation Approach

By the equivalence relation approach, whether an environment is regarded as an accurate simulation of another depends on whether the two environments are equivalent according to a certain criterion. Such a criterion is represented in the form of an equivalence relation, which is a formalisation of the notion of *approximate* and *like*, and it is well understood in mathematics. In this approach a simulation is either 'accurate' or 'not accurate'. Formally a equivalence relation on a set is binary relation which satisfies a set of conditions as described in [33].

By this approach, to answer whether two environments match each other, the criterion of accurate simulation must be defined as an equivalence relation on the software environment. For example, the type equivalence of the environments, i.e. two environments have the same type, is such a relation.

The Accuracy of Real Number System Represented on Digital Computers

Consider a software environment that contains a sensor. The values of the sensor sent to the software are real numbers. Due to the internal representation of the real numbers as floating-point numbers, it is impossible for the software to deal with real numbers precisely. In fact, the application problem may only require that the real numbers are stored and calculated to a certain degree of accuracy. Therefore, the real numbers are divided into equivalent classes such that each class represents the approximate value of the real numbers in the class. For example, assume that the value is required to be held with a precision of 10^{-4}. Then, 1.23456 can be considered as equivalent to 1.2345. To simulate such an environment, a digital computer program can be used which only differs from the real environment in sending the approximate values instead of the exact values to the software. This simulator should be considered as accurate in the sense that the approximate values are equivalent to the exact values for the purposes of the software.

Equivalence in Non-Determinism

Consider an interface between a system and its environment that has two communication channels C_a and C_b. Channel C_a sends data x to the system and the channel C_b sends data y to the system. The order that the data x and y arrives at the ports is non-deterministic, but it is not significant for the software. We can, then, define that the sequence of events 'send data x through channel C_a, send data y through channel C_b' is equivalent to the sequence of events 'send data y through channel C_b, send data x through channel C_a'. Once this equivalence relation is validated and verified, the simulator of the environment only needs to be able to send data x and y in one particular order, say, first send x through C_a, then send y through channel C_b.

Equivalence in Timing and Time Delay

Consider real-time software that receives data through a communication channel. A datum should arrive at the input port every five seconds. The behaviour of the software depends on the punctuality of the arrivals of the data. But, it is allowed that a datum arrives slightly earlier or later as long as the time gap between two data is larger than two seconds and smaller than six seconds.

Assuming that the time space is a set of natural numbers representing time in seconds then the above specification can be considered as an equivalence relation on the sequences of events as described in [33].

Once this relation is validated and verified through the software, the simulator of the environment only needs to generate the data every five seconds to represent every four, five, six and seven seconds.

6.4.2 Metric Space Approach

By the metric space approach, the criterion of accurate simulation of an environment is defined as a metric function on the environments. The accuracy of a simulator is measured by the metric function. Metric space is a formalisation of the notion of distance between objects, i.e. the measure of similarity. It is also well understood in mathematics. A metric space consists of a set A and a function δ from A×A to the real numbers $[0,\infty]$ that satisfies a set of conditions as described in reference [33].

The following paragraphs discuss how to derive metric spaces on environments.

A metric space is a formalisation of the notion of distances between objects. The larger the distance is, the less similar the objects are. In many cases, there exists a measure of the difference of the events. For example, the difference between two events that pass real values through an I/O channel can be measured by the difference of the values. This measure is in fact a metric on the event space. However, from a given metric on the event space, there are many metrics that can be defined on environments. The notion of 'distance' between two instances of an environment needs to be defined first. The basic idea is to measure the 'distance' between two instances by the distances between the states of the two instances at various points of time. The following are some examples.

It is assumed that there is a metric δ (delta) on the event space. Let x, y be two instances of a software environment. The two environments x and y are functions of time. The difference of the two instances at time t can be described as $\delta(x(t), y(t))$, a metric function on the environments.

Maximum Difference Metric

The simplest metric that can be defined on the environments is the maximum of the differences on the events. The maximum difference between the two environments x and y is the least upper bound value of the metric function $\delta(x(t), y(t))$ for a value of t in the time indexed set.

The maximum difference metric is applicable to the situation where for each event the difference has significant influence on the behaviour of the system. However, in practice, not only the difference between each event, but also the accumulation of the differences is important. Hence, the following more general metrics on software environments are useful.

Accumulation Metrics

Consider a software system running on a train. It calculates the distance the train has travelled to determine the precise location of the train. It reads the speed of the train, say every second, from a speedometer. The error in the simulation of the speedometer will be accumulated by the software in the calculation of the travel distance. Hence, to assess the accuracy of the simulation, we need to considered the amount of error accumulated in a period of time.

Since the importance of the difference may vary as time increases, there is a weight function associated with the differences. Let this weight be a function $\omega(t)$, which has a value between 0 and 1. Then the accumulated differences between the two instances of the environment for discrete time is simply the summation of the difference values $\delta(x(t), y(t))$ multiplied by the weight function $\omega(t)$ for each time index.

For continuous time space the accumulated difference is the integral of multiplication of the two functions $\delta(x(t), y(t))$ and $\omega(t)$ between *start time* and infinity. A complete mathematical description of accumulation metrics is provided in [33].

6.4.3 Procedures For Accuracy Evaluation

The following paragraphs discuss how to apply the theory developed in the previous two subsections to evaluate the accuracy of environment simulation. Procedures of accuracy evaluation and assessment are proposed.

Procedure for Assessment of Simulation by Equivalence Relations

The procedure of accuracy assessment according to equivalence relations consists of seven steps illustrated in Figure 6-2.

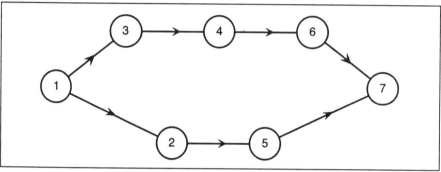

Figure 6-2 : Procedure of Assessing the Accuracy of Environment Simulation

(1) *Identification of equivalent events* There may be a large number of interactions between a piece of software and its environment. It is often impossible to test all the events and their combinations through environment simulation. In fact, it is usually unnecessary because there may exist certain sets of events to which the software always responds in the same way. To reduce the size of event space, the first step is, then, to identify such sets of events. The result is an equivalence relation defined on the event space.

(2) *Validation and verification of the software for correct behaviour on the equivalent events* The equivalence relation approach to accuracy evaluation is based on the equivalence relation defined on the event space or event sequences. The correctness of the equivalence relation is the foundation of whole assessment. Here, correctness means that for all the equivalent events *a* and *b*, the software always responds in the same way and the responses are correct with respect to the specification of the software. Since the behaviour of the software may depend on the history and its internal state, the correctness must be proved in all such contexts. The validation and verification of the correctness of the equivalence relation is the activity of this step.

(3) *Identification of equivalent segments* A large number of combinations of events can be reduced by steps (1) and (2). However, further reduction of the combinations is sometimes necessary and possible because some

segments of the instances of an environment may be equivalent. Therefore, an equivalence relation should be identified and defined on the segments.

(4) *Validation and verification of the software for correct behaviour on the equivalent segments* Similar to step (2), the equivalence relation on the segments must be proved to be correct. Here, the correctness means that:
- for all equivalent segments a and b, the software behaves in the same way and correctly with respect to the specification of the software;
- the whole operation history can always be decomposed into a series of segments;
- the equivalence relation on the segments holds in any context (i.e. the history of the operation and the internal states).

(5) *Verification of the completeness and soundness of event space* When an equivalence relation on the event space is given, the simulator of the environment must be proved to be complete and sound for the event space. Here, completeness means that for any event a in the event space of the real environment, there must be an event a' in the simulation that is equivalent to a. Soundness means that all events of the simulation are valid events, i.e. equivalent to some event in the real environment.

(6) *Verification of the completeness and soundness of event segments* When an equivalence relation on the segments is defined, the simulation must be proved to be complete and sound with respect to the relation. That is, for any segment in the real environment, there is at least one equivalent segment that the simulation may perform, but nothing more. This is again a necessary condition of correctness.

(7) *Validation and verification of the simulation for equivalence to the real environment* The last step, and the most important step, is to show that for any instance of the real environment there is an equivalent instance of the simulator. This can only be done when the real environment is well understood. However, in practice, only a few examples of the real environment instances may be available. In this case, it is necessary to show that all such examples can be simulated.

It is obvious that the more that is known about the real environment, the more accurate the evaluation can be.

Procedure for Measuring Accuracy by Metrics

The measurement of simulation accuracy by metrics could be much more complicated than the assessment of the accuracy by equivalence relations. To reduce the complexity of the measurement, the measurement procedure should start after the assessment by equivalence relations. The measurement procedure consists of the following steps.

(1) *Definition of the metric on the event space*   This is the process of formalising the intuitive notion of differences between events and expressing the notion in a metric function. The general rules are:
- The less different the two events are, the smaller the distance which should be assigned.
- The distance between two equivalent events should be defined to be zero.
- It is advisable to list the events in the order that adjacent events have the smallest differences, and then assign a number as the distance between the adjacent events. The distance between any other two events, say a and b, can be assigned by summing up the distances between the adjacent events in the list between a and b.
- The development of the definition of the metric should go together with the determination of the accuracy target, so that intolerable differences between events are assigned a distance larger than the accuracy target.

(2) *Validation of the metric on event space*   There are no general hard and fast rules for the correctness of the definition of metrics on the event space. It depends on the application domain. Engineering judgement may be necessary for the definition of the metric.

(3) *Definition of the metric on the environment instances*   Once the metric on the event space is defined, the metric on the instances of the environment can be defined. According to the application, either the maximum metric or the accumulation metric can be used. The definition of the metric should be developed together with a target of the accuracy for the simulation.

(4) *Validation of the metric on instances*   Once the metric on the instances of environment is defined, it must be validated according to the application domain. The only potential general rule is that intolerable difference instances must have their distance larger than the accuracy target

(5) *Measuring the accuracy of simulation*   The measurement of the accuracy of simulation is the process of application of the construction of the metric on the software environments. This requires knowledge of the real environment as well as the simulation. However, in many cases, only partial knowledge of the real environment is available, i.e. we can only obtain a subset of the instances of the real environment. Intuitively, the more knowledge of the environment is available (i.e. the larger the subset is), the more accurate the measurement can be.

The measurement process is a repetition of the following activities for each instance of the real environment.
- For each instance of the simulation, calculate its distance from the instance of real environment.

- Find the minimum of the distances calculated above. This is the distance between the instance of the real environment and the simulation.
- Find the maximum of the distances between the instances of the real environment and the simulation. This is the result of the accuracy measurement.

This process of measurement requires a larger number of calculations. An alternative process for the measurement is a repetition of the following activities for each instance of the real environment. This process requires less calculation, but may result in a less useful measurement.

- Identify the instance of the simulation which is 'most similar' to the instance of the real environment.
- Calculate the distance between the two instances according to the metric.
- Find the maximum distances calculated in the previous step over all instances. This is the result of the accuracy measurement.

Probabilistic environments would naturally generalise the problem of assessing simulation accuracy by weighting the contribution of environment instances according to the probabilities of occurrence. This is not discussed further. In practice the issue of accuracy can be treated separately from probability issues. Once sufficient accuracy of simulation instances has been established their probability of occurrence can be specified as a separate problem.

6.5 Justification of Safety from Environment Simulation

In some cases, environment simulation may be used as the basis for closed loop dynamic testing, but forms no part of the safety case used for certification of the system. In others, data from tests performed using this simulation may provide a vital part of the safety case. Clearly, these differing uses to which the results may be put place varying constraints on the development and use of the simulator. If environment simulators are used to validate a system as part of the safety case, then the following issues must be raised at the outset of the simulator development. These issues must be answered before the development of a simulator can begin.

1. How much of the deliverable system is it necessary for the simulator to exercise? For example, can low level I/O be bypassed?

   When a system is developed by a consortium of companies it is usual practice for each consortium member to test their subsystem in a simulated environment; some part of this simulated environment may contain modules of other subsystems. For instance, if the subsystem is software based, then it is recommended that the consortium should design and build the interface packages at the earliest opportunity. This would allow the participating companies to reuse these interface packages to test their subsystem in an environment that is compatible with the other subsystem.

2   What level of complexity is required of the simulator?

The complexity of the simulator is dependent on a number of parameters including the safety criticality level of the system, the complexity of the actual environment, and the functionality and expected use of the simulator.

3   Are there previously validated models/simulators available?

If a simulator is to be used for validation purposes then the reusability of an existing simulator must be considered. For instance, how much of the existing simulator can be reused? Surveys show that most companies developing safety-related systems usually produce systems of only one kind with an identical environment. If, for example, the new system is an upgrade of an existing system the addition of more processing power, a better user interface, or better data logging facilities, then typically about 80 percent of the existing simulator will be reusable.

Safety standards [13] recommend that if a simulator is used to validate a safety-related system then the simulator should be properly validated. In industry, simulators are generally validated using ad hoc techniques and no guidelines on simulator validation are available. Research in validation techniques of simulation models has produced a number of major practical validation techniques ([34], [35], [36] and [37]). One process of simulator validation is discussed below.

In Figure 6-3, the text in the rectangles represent states of the simulator development and the solid arrows represent the evolution of the development of the simulator. The dashed lines show where validation and verification are performed. If the simulation results provide a vital part of the safety case then a peer assessment method can be used to evaluate the acceptability of the simulation results. This panel of peers should be composed of:

- engineers who have expert knowledge of the system under study, and who know the objectives of the environment simulation, e.g. the project manager or chief engineer
- experienced system engineers
- experienced software and hardware engineers
- engineers with extensive experience with simulation projects
- quality assurance staff within the organisation
- experienced safety assessors.

Simulators are based on abstract models of real world systems. Confidence in the model requires not only knowledge about the model, but information about the extrapolation involved in dealing with specific problems. In any abstract model of a system certain assumptions are made and the validity of these assumptions is crucial to the validation of the simulator.

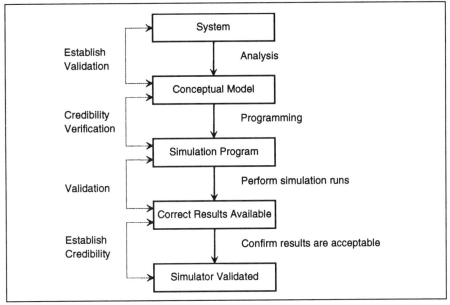

Figure 6-3: Simulator Verification and Validation Process

6.6 Guidance

> **Guidance on the Use of Simulators**
>
> - Developers should select an appropriate model for their environment simulator. If the simulator models a deterministic environment then a set theoretical model is appropriate, if the environment is non-deterministic then a stochastic model should be used.
>
> - Developers should consider quantifying the accuracy of the simulation result using one of the approaches described in this chapter.
>
> - The safety case documentation should include sufficient evidence of the verification and validation processes (including accuracy checks) of the simulation.
>
> - If the simulation results form a part of the safety case argument then the simulator should be validated using guidelines recommended in the industry specific standards.

Chapter 7
Test Adequacy

The specification of test adequacy is important as it sets limits on the amount of testing that is judged to be sufficient. It also sets a level of confidence that can be associated with the testing. Different approaches to measures of test adequacy are discussed, and guidelines on the use of test adequacy criteria are presented.

7.1 Introduction

In 1975, in the examination of the capability of testing for demonstrating the absence of errors in a program, Goodenough and Gerhart [38] made a significant break-through in the research in software testing by pointing out that the central question to be addressed in software testing is 'what is a test criterion?', i.e. the criterion which defines what constitutes an adequate test. Since then, test criteria have been the research focus of software testing. The past two decades have seen a rapid growth of interest in software test criteria. Many such criteria have been proposed and investigated. A huge amount of research and many experiments have been conducted to provide various rationales to support the use of one criterion or another. Research results indicate that different adequacy criteria may have different fault detecting abilities. The confidence in the quality of the software which is derived from the software passing an adequate test may also vary with adequacy criteria. This chapter surveys test adequacy criteria and discusses their uses in the testing of safety-related software.

7.2 The Notion of Test Adequacy

From Goodenough and Gerhart's point of view [38], a software test adequacy criterion is a predicate which defines 'what properties of a program must be exercised to constitute a *thorough* test, i.e. one whose successful execution implies no errors in a tested program.' To guarantee a tested program to be errorless, they proposed two requirements of test criteria, namely, reliability and validity. Reliability requires that a test criterion always produces consistent test results, i.e. if the program is tested successfully on one test set which satisfies the criterion, then on all test sets which satisfy the criterion the program is also tested successfully. Validity requires that the test always produces meaningful results, i.e. for every error in a program, there exists a test set which satisfies the criterion and is capable of revealing the error.

However, it was soon recognised that these two requirements, especially the reliability requirement, were too strong to define practically useful test adequacy criteria. Since then, the focus on test adequacy criteria seems have been transferred to the search for practically applicable approximations to the ideal criteria.

Subsequently, the software testing literature contains two different, but closely related, notions associated with the term *test data adequacy criteria*. Firstly, an adequacy criterion may be considered as a stop rule which determines whether enough testing has been done. For instance, a test set may be considered as adequate if the execution of the program on the test set exercises all the statements of the program. In this sense, 'exercising all statements' is a test data adequacy criterion.

Secondly, test data adequacy criteria may be measures of the quality of testing, that is, they measure the ability of a test set to reveal a particular feature of a program. Hence, a degree of adequacy is associated with each test set. In this sense, the coverage of code as a percentage is often used as an adequacy measure. Here, test sets are not simply classified as either good or bad.

These two notions of test data adequacy criteria are closely related. A stop rule may be considered to be a special case of a continuous measure.

7.3 The Role of Test Data Adequacy Criteria

Test data adequacy measurement plays important roles in software testing. Two such roles in the theory of software testing and software test management are discussed below.

7.3.1 The Role in Software Testing Theory

An adequacy criterion is an essential part of any testing method. It plays two fundamental roles in every software testing method: to select the test cases themselves, and to determine which observations must be made during testing.

Firstly, an adequacy criterion can determine what should be used as test cases. There are two ways to use an adequacy criterion to determine the test cases.

- The explicit approach is to specify what kind of test cases are required. For instance, statement coverage requires that an adequate test set exercises all the statements of the program under test. Such a criterion is usually referred to as a test case selection criterion. Using a test case selection criterion, a testing method may be defined constructively in the form of an algorithm which generates a test set from the program under test. This test set is then considered adequate.

- The second way is to measure the adequacy of a given test set generated in some fashion. Examples of such criteria include mutation adequacy and Weyuker's inference adequacy criteria [39]. A rule determines whether a test set is adequate, or more generally, how adequate, and it is this that is usually referred to as an adequacy criterion.

Secondly, an adequacy criterion also determines what observations should be made during software testing. For example, statement coverage requires that the tester, or the testing system, observes whether or not each statement is executed during the process of software testing. Such information must be collected to calculate the overall coverage. If path coverage is used, then observation of whether statements have been executed is insufficient. Path testing requires that the execution paths are recorded. However, if mutation score is used as the adequacy criterion, it is unnecessary to observe whether a statement is executed during testing. Only the output of the original program under test and the output of the mutants must be recorded and compared.

Due to the central role that adequacy criteria play in software testing methods, the comparison of software testing methods is often performed by comparing the test data adequacy criteria underlying the methods. Although given an adequacy criterion different methods could be developed to generate test sets automatically or to select test cases systematically and efficiently, the quality of a test method for software quality assurance purposes is largely determined by the adequacy criterion. In addition, it can also determine the efficiency of the testing. Unfortunately, it is not yet clear what is the exact relationship between a particular adequacy criterion and the correctness or reliability of software which passes the test. But, it is certainly true that the quality of the software cannot be assured if the testing is inadequate according to very basic criteria such as statement coverage and branch coverage.

The term 'adequacy criterion' is used as a synonym for 'testing method' when there is no possibility of confusion.

The notion of adequacy criteria is a cornerstone of software testing theory. Based on it other notions are built. For example, the notion of software testability can be defined as how easily an adequate test can be performed. For example, Bache and Mullerburg [32] defined software testability to be the minimum number of test cases that satisfy an adequacy criterion. Moreover, the complexity of testing is the relationship between the size of program and the size of an adequate test set. This can only be studied with respect to a given adequacy criterion. Finally, test case selection criteria are closely related to test adequacy criteria, as seen above. In many cases they can be easily transformed from one form to the another.

7.3.2 The Role in Software Test Management

One of the most important issues in the management of software testing is to 'ensure that before any testing the objectives of that testing are known and agreed and that the objectives are set in terms that can be measured', Such objectives 'should be quantified, reasonable and achievable'[40]. Test adequacy criteria are objective rules applicable by project managers for this purpose. For example, a structure coverage criterion, such as branch coverage, is used to justify the quality of a test according to whether all the structures concerned, such as branches in the program, have been exercised by testing. Mutation adequacy justifies the quality by checking if all possible faults, in some sense, have been found by testing. The quality of functional testing is justified by examining if all the functions of the software have been tested.

Test data adequacy criteria are also very helpful tools for software testers. From a practical point of view, there are two different levels of software testing processes. At a lower level, testing is a process where a program is tested by feeding more and more test cases to the program. Here, test data adequacy criteria can be used as a stop rule to decide when this testing procedure has achieved the objectives and can be stopped. This role of adequacy criteria has been considered to be the most important one by some computer scientists. Moreover, when the measurement of test adequacy shows that a test has not achieved the objectives, adequacy criteria also provide guidelines for the selection of additional test cases.

At a higher level, a testing procedure can be considered as repeated cycles of testing, debugging, modifying program code, and then testing again. Here, testing can be stopped when the software has met a given reliability requirement and known errors in the program can sometimes be ignored or accepted. Although test data adequacy criteria do not provide the stop rules at this level, they make an important contribution to the assessment of software reliability. Generally speaking, there are two basic aspects of software reliability assessment. One is the estimation of the reliability figure, the other is the estimation of the confidence or accuracy of the reliability estimation. The role of test adequacy here is as a contributory factor in building confidence in the accuracy of a reliability estimate.

Therefore, adequacy criteria enable testers to manage software testing procedures so that the quality of the software can be assured by performing sufficient tests. Once adequacy has been assured, there is no need to perform further testing and so proper use of adequacy criteria also enables software developers to control the cost of testing by avoiding redundant and unnecessary tests.

It is widely recognised that human factors play a significant role in the effectiveness of software testing. For example, Basili and Selby [41] noted that software testers' professional experience significantly affects their ability for fault detection when using various software testing methods. Myers [42] also found high variability both among the number of faults detected and the types of fault detected even among highly experienced professionals. One way to control the influence of human factors in the effectiveness of software testing is by using very stringent adequacy criteria so that the testing can only be performed very rigorously.

7.4 Approaches to Measurement of Software Test Adequacy

There are various ways to classify adequacy criteria and a wide variety of terms must be understood in order to follow such classifications. This section explains those terms briefly. Full explanations and analysis can be found in Zhu's report entitled *Software Testing Adequacy and Coverage* [43].

7.4.1 Specification, Program and Usage Based Criteria

One of the most common classification methods is based on the source of information used in the measurement of adequacy.

In white box testing, program based adequacy criteria are used. These indicate whether a test set is adequate or measure the adequacy of a test set by analysing information gained from the program under test. This class of adequacy criteria has dominated the research in software testing. In black box testing, test adequacy is decided according to the information available in the specification. Such adequacy criteria belong to the category of specification based adequacy criteria.

There are three major types of information contained in specifications which can be utilised in test adequacy measurement.

Required Functions A specification contains information about the functions to be implemented by the software. These functions, usually called specification functions, together with functions introduced during the design and implementation of the software, which are called design functions and implementation functions, are the main objects of functional analysis and functional testing.

Input/Output Domains The second type of information contained in specifications is that of the input and output spaces. This information can be used in boundary analysis by dividing the input and output spaces into several subspaces and checking that test data have been selected in the middle, on the boundaries or near the boundaries for each subspace.

Syntactic Structure Finally, the syntactic structure/coverage of a specification also provides important information for test adequacy measurement. The correctness of the output of a program must be checked according to the specification. This might involve the 'execution' of the specification in some sense. Hence, the adequacy of a test set can be analysed by the coverage of the syntactic structures being used in the executions. Such adequacy criteria belong to the syntactic coverage category.

It has been widely acknowledged that a better software testing practice should use information from both the specification and program. However, there are few such adequacy criteria proposed and investigated. Most research into specification based test adequacy criteria extends testing methods first developed as program based testing.

An adequacy criterion can also be based on the information about the usage of the software such as the probability distribution on input data space. This information is used in random testing.

Hence, most existing adequacy criteria are either specification based, program based, both specification and program based, or usage based.

7.4.2 Testing Based upon Adequacy Criteria

Another approach is to classify test adequacy criteria according to the underlying testing approach. There are three basic approaches to software testing. The first is the structure coverage approach. The basic idea is that an adequate test should exercise all or most of the structure of the program (or the specification) under test. The second testing method is error based testing, which requires test cases to check the program on those error-prone points according to our previous empirical knowledge about software errors. The third approach to software testing is fault based testing, which focuses on the detection of faults in the software.

Structure Coverage Measures A structure coverage measure uses information about the execution behaviour of the program (or executable specification) during testing. It measures adequacy by calculating the proportion of a certain set of elements or structures exercised by testing. Such elements or structures can be defined either in terms of a model of the program structure or directly using the program text. In the latter case, the criterion is called program text based, while in the former case, a flow graph model of the program structure is often used, and the criterion is called flow graph based. This subclass is then further divided into control flow criteria and data flow criteria according to the kinds of paths concerned.

When programs are modelled as directed graphs, paths in the flow graph should be exercised by testing. However, due to finite computing resources, only a finite subset of the paths can be checked during testing. The problem is therefore to decide which paths should be exercised.

Control flow test data adequacy criteria answer this question by specifying either restrictions on the complexity of paths or the redundancy among the paths. However, without further semantic information, it is hard to see how the importance of a path is related to the semantics and the use of the program. Data flow test data adequacy criteria attempt to associate additional semantic information with flow graphs – data flow information– and to select paths which are significant with respect to such information.

Data flow adequacy criteria make more use of program semantics information. Consequently, test data that satisfy such criteria can be more effective than those that only satisfy control flow criteria. However, it is important to understand that satisfaction of data flow criteria and control flow criteria does not require any reference to the required functions of the software under test.

Both control flow and data flow based test data adequacy criteria use a flow graph model of program structure. To apply such adequacy criteria, a flow graph must be created from the program text. A more direct approach is to use the program text. The advantage of text based adequacy criteria is that the criteria can easily be related to the program text. However, for programs written in structured programming languages, reformatting and static analysis of the program are necessary for the application of non-trivial adequacy criteria. In such cases, the connection between program text and the criteria becomes less straightforward.

Error Based Criteria Program error based criteria analyse test data adequacy according to whether certain kinds of errors have been checked by a test set. The basic idea behind the approach is the classification of program errors into two types: computation errors and domain errors. A computation error is reflected by an incorrect function computed by the program. Such an error may be caused, for example, by the execution of an inappropriate assignment statement that affects the function computed by a path in the program. A domain error may occur, for instance, when a branch predicate is expressed incorrectly or an assignment statement that affects a branch predicate is wrong, thus affecting the conditions under which the path is selected. Boundary analysis adequacy criteria focus on the correctness of the boundaries, where generally domain errors occur, and so are biased to find domain errors. Functional analysis criteria focus on the correctness of the computation, and are biased to find computation errors. Both types of criteria have their own weaknesses, hence they should be used to complement each other.

It is widely recognised that software testing should take both the specification and the program into account. A way to combine program based and specification based domain analysis testing techniques is first to partition the input space using the two methods separately, then to refine the partition by intersection of the subdomains. Finally, for each subdomain in the refined partition, the required function and the computed function are compared to see if they are the same kind of functions, say polynomials of the same degree, and the test cases are selected according to the set of functions they belong to.

A limitation of domain analysis techniques is that their application is complicated when the software has a complex input space. For example, process control software may have sequences of interactions between the software and the environment system in the behaviour space. It may be difficult to partition the input space into subdomains.

Another shortcoming of boundary analysis techniques is that they are defined on some informal notions, such as the closeness of two points and 'just off a border'. These notions can be more or less formalised for a numerical input space, but it is not a simple problem for non-numerical software, such as compilers.

In the identification of functions for functional analysis, it is suggested that not only specification functions, but also design functions and implementation functions (i.e. the functions introduced during the design and implementation phase in software development) should be tested.

Fault Based Criteria Program fault based adequacy criteria measure test adequacy according the ability of a test set to detect faults in a program. They focus on the possible faults introduced into the software. The adequacy of a test set is measured according to the ability of the test set to detect such faults. Typical examples of fault based adequacy criteria are error seeding and mutation adequacy.

Error seeding plants artificial faults into the software and uses the ratio of the number of the errors found by testing over the total number of seeded errors as an adequacy measure. It is based on the assumption that the artificial faults planted into the program are as difficult to detect as the natural errors. This assumption has proved untrue in general.

Mutation analysis systematically and automatically generates mutants of the program under test. Each mutant contains one fault. The test adequacy is measured according to the ratio of the number of mutants found to be different from the original program over the total number of non-equivalent mutants. It is powerful, and has a relatively sound theoretical foundation. The principle of mutation adequacy analysis can easily be extended to specification based adequacy analysis. The most important issue in mutation adequacy analysis is the design of mutation operators, i.e. a similar problem applies to that of error seeding, in that the faults resulting from mutation may not be similar to 'natural' faults.

Measuring the adequacy of software testing by mutation analysis is expensive, and may require a huge amount of computation resource. How to reduce the testing expense to a practically acceptable level has been an active research topic. Variants such as weak mutation testing, firm mutation testing and ordered mutation testing have been proposed.

7.4.3 The Space of Adequacy Criteria

Based on the above discussion, software test adequacy criteria can be put into a multi-dimensional space, as shown below in Figure 7-1. Each approach to the measurement of software test adequacy is represented by an axis in the diagram. An adequacy criterion combining two approaches is then a point in the two dimensional plane formed by the corresponding axes. For example, a program mutation adequacy criterion is a program based and fault based criterion. Hence, it is on the plane formed by the axes of program based criteria and fault based criteria. Similarly, if an adequacy criterion combines three approaches to the measurement of software testing adequacy, it is in the three dimensional space formed by the three axes corresponding to the three approaches.

An adequacy criterion can also be based upon information about the usage of software, such as the probability distribution on the input space, which plays an important role in random or statistical software testing. The subject of random testing is treated in detail in Chapter 8 which relates the notion of test adequacy to statistical software testing, and argues that the reliability estimates which result from such testing can be interpreted as an adequacy measure. Chapter 8 also notes the value of the fusion of the topics of test adequacy measures and statistical software testing by way of the role that the

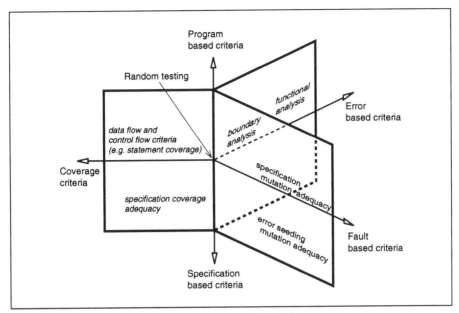

Figure 7-1 : Categories of Test Data Adequacy Criteria

conventional notions of test adequacy, as treated within this chapter, have within improved statistical testing techniques.

Figure 7-2 gives a list of the most well known test data adequacy criteria, presented in the form of a hierarchy structure consisting of classes, and subclasses. Readers are referred to Zhu's report entitled *Software Testing Adequacy and Coverage*, [43] for the definitions of these criteria.

This chapter has focused upon conventional measures of test adequacy. However, adequacy may also be considered from more general or unconventional perspectives and such treatment of adequacy related topics may be found elsewhere in this book. As noted above, Chapter 8 considers the notion of adequacy in a wider sense from the perspective of random testing. Chapter 9 also considers general measures of the achievement of adequacy and integrity in quantitative terms through consideration of testability and test regime assessment. Finally, the adequacy of any test will be dependant upon how representative that test is of the actual demands which the software will experience in its operational environment. This issue is addressed under the topic of accuracy of environment simulation, which has been discussed in Chapter 6, as well as under the topic of random testing in Chapter 8.

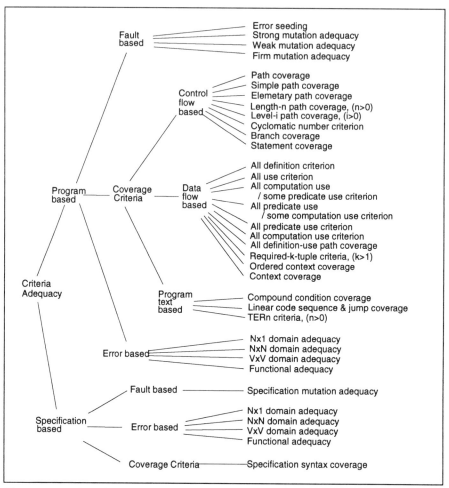

Figure 7-2: Classification of Test Data Adequacy Criteria.

7.5 The Use of Test Data Adequacy

There are a wide variety of rationales presented to support the use of one or another criterion, yet there is no clear consensus. Such rationales fall into three main classes.

The first kind uses statistical or empirical data to compare the effectiveness of criteria. A typical example of this kind is Duran and Ntafos' work [44], which used simulation to compare the probabilities that random testing and partition testing methods would detect errors.

The second analytically compares test data adequacy based on certain mathematical models of software testing. Gourlay's [45] mathematical framework of software testing belongs to this class.

The third studies the abstract properties of test data adequacy criteria and assesses adequacy criteria against axioms representing our intuition about software adequacy. It includes Weyuker's work [46], Parrish and Zweben's formalisation and refinement of Weyuker's axiom system [47], [48], and Zhu, Hall and May's work based on measurement theory [49], [50].

The majority of analytic comparisons in the literature use the subsume ordering, where a criterion A subsumes criterion B if for any program P under test, any specification S and any test sets T, T is adequate according to A implies that T is adequate according to B. A relatively complete picture of the subsume relations between test adequacy criteria has been built. Figure 7-3 is a summary of the results which can be found in the literature ([51], [52] and [53]).

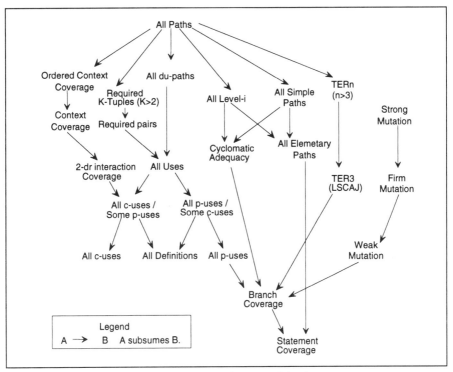

Figure 7-3 : Subsume Relation Between Adequacy Criteria

However, only those criteria which are based on the same model can be compared using subsumes. Many methods are incomparable, i.e. neither subsumes the other. Recently, Frankl and Weyuker [54] proved that a criterion A subsumes criterion B does not guarantee that test data satisfying A is more likely to detect faults then test data which satisfy criterion B. Considering the ability to detect errors in a program to be one of the most required features of adequacy criteria, Frankl and Weyuker proposed some relations between adequacy criteria which are proved to be more directly related to the ability of error detection. However, subsume relation still gives a useful indication of the strictness of the testing that a criterion requires. Therefore, the following guidelines for the use of test data adequacy criteria are put forward.

The principle behind these guidelines is that when choosing test data adequacy criteria for the measurement of the quality of software dynamic testing, both the required integrity level and features of the software, such as complexity, should be taken into account. Because each criterion has its own strength and weakness, more than one complementary criterion should be applied.

7.6 Guidance

- It is important to specify adequacy and coverage criteria as part of software test planning at an early stage in the development cycle.
- Higher integrity level require stricter adequacy criteria. Figure 7-3 can be used as a guideline for the strictness of test adequacy criteria.
- More complex software requires stricter adequacy criteria. In particular,
  - The more complicated the control flow structure of the software, the stricter the control flow based adequacy criteria should be.
  - The more complicated the data flow structure of the software, the stricter the data flow criteria should be.
- Use of more than one test data adequacy criteria from different classes can be more accurate for test quality measurement and more efficient for software test than use of only one extremely strict adequacy criterion.
- Testing should always aim to achieve a high degree of test adequacy. But, if the maximum or almost maximum degree of adequacy is achieved, it may also indicate that the criterion in use is not strict enough. It is important to be aware that adequacy criteria can only show the weakness of the test. High degree of adequacy under a criterion of low strictness does not always mean high test quality.
- A high degree of adequacy for the whole program is not sufficient, an even distribution of the adequacy over the components of the software is also required. In particular, a low degree of test adequacy for a critical part of the software is not acceptable.

Chapter 8
Statistical Software Testing

This chapter discusses the statistical testing of software. Statistical Software Testing (SST) is gaining more acceptance as a method of assessing software integrity and provides a quantifiable measure of reliability. The tasks necessary to carry out SST are described. SST for risk estimation is compared against SST for failure probability estimation. Software environment simulation for SST is particularly demanding; the various issues are described. SST is related to test adequacy measurement as commonly understood.

8.1 Introduction

Statistical software testing, which is often referred to as 'random testing', is now an established form of software testing. The great benefit of statistical software testing (SST) is that it provides a directly quantifiable metric for perhaps the most sought after measures of software integrity, namely, the probability of failure and the closely related measure of reliability [55]. Such measures are sought after because they can be employed in existing forms of quantified safety arguments for systems including software. Standards for safety-related systems often specify required integrities for systems of a certain criticality in terms of these measures [6] [13]. Currently, statistically based methods are the only methods which can directly justify integrity in this way.

Statistical software testing typically uses environment simulation to exercise the software. There are currently many issues in need of clarification when simulation is used for this task. For example, what constitutes a single 'test?' How can test results be observed? Can the success of a test be automatically decided (i.e. how complex does the 'oracle' need to be in practice)? The simulation of software environments is a large and complicated task and the results will emerge as the subject area develops. This chapter distinguishes two different types of software environment simulation. The key factor discriminating these two types of environmental simulation is the degree of interaction between the software and its environment. 'Pure protection system' software does not influence its environment during the period in which it decides whether it will take action. In contrast, control software exerts an influence on its environment. These differences lead to some different experimental techniques when testing via environment simulation.

8.2 Statistical Software Testing and Related Work

Currently, there are three fundamentally different theoretical approaches to statistical software integrity prediction: these are called *single-version testing, reliability growth modelling* and *decompositional modelling*.

Single-Version Testing

As its name implies, a single-version testing technique tests a single unchanging piece of software code, and infers the probability of failure from the test results. Two different types of statistical model can be identified within the category of single-version techniques.

- *Time based models*, in which the emphasis is on the density of failures over time [56], and where estimation is based on the observed times between successive failures (or in the case where no failures have been observed, the length of time for which this has been the case).

- *Demand based models*, in which the emphasis is on the distribution of the number of test failures in a number of distinct 'tests,' known as demands, and where estimation is based on the number of failed demands in a sample of demands randomly selected from the operational input distribution of the software.

In the remainder of the chapter, SST means single-version testing. As explained later, demand based SST applies in an obvious way to pure protection systems, but can be applied to any system. This chapter focuses on demand based SST. To date all statistics for demand based SST have been based on an application of urn models, and consequently the binomial probability model of failures. Such models have a long pedigree, having been originally derived by Laplace. Works which use this basic model include [55], [57], [58], [59] and [60].

Reliability Growth Modelling

The second approach is based on the study of successive releases of a software product. This approach is often called the study of reliability growth [61]. Successive releases are usually considered to arise from the debugging cycle. Mills et al. [62] propose an alternative view where successive releases arise as extra functionality is added, in a software lifecycle based on stepwise refinement. In either case, reliability growth models must employ assumptions regarding the effects on integrity, of making alterations to code. Reliability growth models are not considered in detail here precisely because integration of new code with old code can have effects of a very unpredictable nature – the possibilities appear to be unlimited – and so general assumptions to describe these effects are likely to fail in some circumstances, as in fact they do sometimes fail [63]. It is difficult to justify measurement of integrity for safety-related systems based on such assumptions.

The relationships between the different techniques of reliability growth and single-version integrity estimation is not well understood. In a paper on 'cleanroom techniques' both techniques are considered [59], and it is noted that more testing is required to justify a given level of integrity when using the single-version approach. Clearly this could either be because the single-version techniques are conservative or because the reliability growth techniques are optimistic.

Decompositional Modelling

A third strand of work which is orthogonal to the above two approaches employs software decomposition. The software integrity is estimated from estimates of software component integrities, which are combined using suitable probability models. These can be called decompositional models, and examples of work in this area for software can be found in [64], [59], [65] and [66]. Clearly, the integrities of the components must be estimated using either the first or second approaches above. The field of decompositional models remains controversial with slow progress due to the different nature of software faults as compared to hardware faults, for which decomposition models are quite mature. Technically the problem lies in modelling the dependencies between software components. There are no comprehensive models of these dependencies currently available. It appears that a component can depend on others in almost arbitrarily complex ways. There is some interesting foundation work in this field. For multi-version software it has been shown that even if development is 'independent' in a well-defined formal sense, the failure behaviour can be dependent. A similar observation has been made experimentally [67]. Of course it might be expected intuitively that two codes performing the same task might exhibit dependent failure behaviour. However, the dependence problem is not restricted to such codes [68].

8.3 Test Adequacy and Statistical Software Testing

The notions of test adequacy and software integrity appear close. However, they are not the same concept, and as a result the literature of the two areas is currently fairly disjoint. SST is usually discussed as a means of measuring software integrity, so what does the notion of adequacy mean in relation to SST? In the discussion below, it is argued that the reliability estimates provided by SST which can be interpreted as 'special' kinds of adequacy measure.

In the general sense, the integrity of software should be a measure of its nearness to correctness. The most widely used attempts to measure integrity are reliability estimates, and probability of failure estimates, which are closely related to each other. A program has an actual or 'true' reliability or probability of failure, which is a measure of the degree to which the program fulfils its required function, and it is this which the statistical techniques estimate. The information contained in statements of reliability and probability of failure are sufficient for the purposes of safety arguments, and hence their significance.

The idea of test adequacy is slightly different. It is a property of a test set, a program, and a specification. Perhaps the most obvious contrast between adequacy and integrity is that integrity depends on the results of the testing whereas adequacy is usually described as a property independent of test results.

There are many candidate measures for adequacy [49]. The control path and data path coverage measures are the best known. Interpretation of these common adequacy measures is problematic. Despite their quantitative nature, there appears to be no quantitative way of using such coverage measures in a safety argument. It must be remembered what adequacy really attempts to measure, and we suggest that this is the ability of a test set to demonstrate that no faults are present in the code under test should all tests succeed. The code could be thought of as possessing a 'fault propensity,' a potential for faults to be remaining within it. As more successful tests are made the potential for remaining faults is reduced. The adequacy to which a program has been tested (i.e. the adequacy of the set comprising all tests conducted) should be the antithesis of this fault propensity. More tests can be equivalently viewed as resulting in a reduction of fault propensity; or an increase in test adequacy. Some recent work on adequacy is consistent with this view i.e. adequacy as the ability to expose faults. It is clear that many proposed adequacy measures cannot properly fulfil the requirements of this view. For example, for none of the control or data path coverage methods does 100% coverage imply the absence of faults; this is even true for 100% code path coverage, since the specification may include a feature which the code does not implement, or the behaviour of a path may depend on the initial program state. Therefore, making use of coverage measures only allows us a limited form of safety claim – for example, we can say with certainty that we would always prefer to add more paths to the list of tested paths, but as we increase the coverage we do not know how much we decrease fault propensity. In fact, none of the coverage measures can describe fault propensity.

The above discussion emphasises the point that, to decide on an adequacy measure, it is necessary to be specific about what information the measure is to provide. For example, it would be possible to choose to measure adequacy in terms of ability to expose faults, with no emphasis on a particular classification of faults. However, another view would choose to measure adequacy as the ability to expose faults commensurate with their likelihood of occurrence during normal operation. In the latter case, but not the former, probability of failure or reliability could be used as the measure. Thus, SST provides a measure which is both an integrity and a particular kind of adequacy, namely, the estimate of probability of failure which would be justified statistically should all tests succeed.

SST does not only provide a single-value estimate of the probability of failure. A confidence in this value is also provided (see 8.5.2). Strictly, it would be more accurate to say that SST produces an adequacy measurement which is a probability distribution, rather than a point value probability. However, the distribution can be summarised using a point value 'best estimate' of the probability of failure (the expected value or mean of the distribution) and a point value describing the spread or confidence in the point value estimate (the variance of the distribution). Both the best estimate and the confidence value may be regarded as adequacy measures. The measure of confidence, whether it be a variance or a confidence interval, varies with the test set. Different test sets produce different probability of failure estimates and different confidences in that estimate. The confidence is an 'adequacy' in the general sense of the word, but it is not a measure of fault propensity and so is not an adequacy in the sense of the word used above in this section.

Whilst the traditional notions of adequacy (statement coverage, path coverage etc.) are not directly usable in quantified safety analyses, some recent research indicates that these notions may have a role to play within improved SST techniques [69]. Such a fusion of topics (traditional test adequacy measures and SST) may be encouraging to some software quality practitioners who have regarded SST with suspicion on account of its complete disregard for issues such as code coverage. After all, the study of coverage occurred because it was felt, intuitively, that coverage was an important factor in demonstrating software quality by testing. The new work is suggesting that this is true, but that coverage is not the only concern.

8.4 Environment Simulations in Dynamic Software Testing

This section raises some general issues regarding environment simulation. The particular significance of environment simulation for SST is considered in Section 8.5.1. The degree of detail in the simulation which is constructed to test a piece of software depends not only on the real software environment but on the software requirements. Firstly, there is the obvious point that the simulation need not include aspects of the environment which neither affect nor are affected by the software. Secondly, the simulation need only be as detailed as the software is able to discern. For example, the sensors returning environment information to a reactor protection system will only be sensitive to discrete intervals in the range of the real world quantities they measure. As long as the simulation places its simulated quantities to the accuracy of these intervals further accuracy is irrelevant.

The production of environment simulators for control software is of particular interest, since the software actually affects the environment behaviour. It follows that the environment behaviour cannot be simulated independently of the software. This situation can be contrasted with that for pure protection system software, where the behaviour of the environment which determines the software behaviour is unaffected by the software and can be modelled independently (it would be possible to run the environment simulation separately to collect the protection system inputs from the environment in a data file, and then use this data file to test the software.) These two cases are referred to as closed and open loop environment-software systems, respectively.

8.5 Performing Statistical Software Testing

To perform SST it is necessary to:
- construct an environment simulation for the software based on the software's operational distribution
- conduct statistical inference from observations of the software's performance to produce an estimate of the probability of failure and a measure of the confidence in that estimate.

8.5.1 Constructing Operational Distributions (Probabilistic Environment Simulations)

To perform SST it is necessary to sample from the operational input distribution of the software. That is, the probabilistic properties (i.e. some events happen more often than others) of the simulated environment which is used to test the software must correspond to those of the real environment in which the software is to operate. This is often a complex, many-faceted task as described below for demand based SST.

In any probability analysis it is necessary to define the events of interest. The fundamental event in demand based SST is a demand on the software. A demand is a single event in the input space of the software. The way in which a single event in the input space is defined depends on those features of the input space which are relevant to whether or not the software behaves correctly. For example, for a program which computed the sine of a single real value, it could be a single datum input. In general, it will be a sequence of inputs, or a timed sequence of inputs.

The concept of a demand is best explained using examples. For a plant protection system which runs in a steady state until the plant leaves its steady state behaviour, one way to define an event in the software input space is as a set of input trajectories i.e. each input parameter to the protection system follows a trajectory, and an event is a set consisting of a trajectory for each input parameter over a time interval. Such an event is called an environment state development or scenario. The space of demands is then the set of environment state developments by which the plant can develop from its steady state to reach a shutdown state i.e. a plant state for which the protection system is required to intervene and shut down the plant. For an aircraft engine control system, the situation is more complex since the control system influences its environment (the system is closed loop). A demand can be defined in a similar fashion to that for a plant protection system. However, it is important to note that the demand results from a development of the state of the software plus simulation system from a steady state (provided such a state exists), and is not simply determined by features within the environment. If the 'software plus environment simulation' system does not have a steady state in operation then a demand consists of one (of many possible) starting state plus a state development.

Demands are based on the idea of a development of state. In order to be relevant evidence for the inference of software probability of failure on demand (pfd), such developments should be restricted to those which can occur in practice. Beyond the requirement of realism, there are two conflicting influences which determine which environmental developments are to be included as demands in an environment simulation.

The Role of Hazard Analysis in Defining the Demand Space

The first influence attempts to restrict the number of environment developments entering the space of simulated demands. It may be the case that the pfd is only required for a subset of all realistic developments i.e. that the behaviour of the software under some real environment state developments is not important within a particular safety argument. This can happen where the consequences of a state development are low, or where the presence of another system such as a hardware plant protection system is sufficient to satisfy safety requirements. *Environment Hazard Analysis* (HA) determines the subset of the input space for which a pfd is required, which therefore forms a part of any strategy of testing via environment simulation. This process requires that SST be viewed in the context of an overall safety argument. HA is performed on the software environment, and identifies those environment transients for which functioning of the software is important. For example, in order to fulfil some overall plant safety levels, a software protection system for a nuclear power station may be required to function with a pfd of 10^{-3} for a large 'loss of coolant' accident, but a 'steam line break' may be a hazard which is rarer or less hazardous and so, for the purposes of the safety argument, may be adequately dealt with by other plant safety features alone.

For a software control system, such as an aircraft engine controller, it may not be possible to ignore any environmental events in this way. This is because such a controller may be required to work for all of the engine mission time i.e. all environmental events. High integrity is required for all possible environment events. However, it remains necessary to identify the relevant set of events from the environment (i.e. the engine and atmosphere in this case). For an aircraft engine, the normal events would not usually be considered hazard-causing but could certainly lead to unsafe system states due to software design faults, and should therefore be tested. These normal events would include steady state flight at various heights and various patterns of aircraft ascent and descent. However, environmental events with more obvious potential for hazard-causing would be mechanical breakages of engine components, objects (e.g. birds) entering the intakes, and violently fluctuating atmospheric conditions.

The Role of Software in Defining the Demand Space

The second influence restricts the extent to which the first influence can be employed. That is, the second influence works to maintain the number and complexity of environment developments which must be included in the space of simulated demands, such that testing is close to using the full operational input space of the software, including its probabilistic properties. This influence is derived from the nature of the software itself, and hence it requires an analysis which is introspective to the software. The second influence arises because all factors relevant to software failure should be included in the simulation, and one of these factors is the initial state of the software for each demand. This is the same issue as asking whether the responses to demands on the software are interrelated in any way. If not, the order and relative frequency of demands can be varied without influencing software behaviour. Interrelated behaviour between demands occurs when previous demands affect the program state entering a subsequent demand, where the value of the transferred state affects the software behaviour and hence could make the difference between success or failure on that subsequent demand. Where this state transfer is important, any attempts to restrict the demand space by according to the 'first influence' above may be negated since, whilst some set of demands D may not be of direct interest, the behaviour of the software for other 'interesting' demands may depend on whether or not demands from D have occurred. In other words, unless demands from D are executed (and according to the operational distribution of all demands), the software may behave in an unrepresentative fashion. For closed environment-software systems the situation is made more complex because any unrepresentative behaviour on the part of the software will feed back to the environment simulator, and the problem takes on an undesirable feedback loop quality.

The role of program state in statistical software testing is a particularly difficult problem. There are two possible approaches:

- argue that the program state is irrelevant, for example the program might possess a steady state which is the initial state for any demand (such a state might be designed into the software)
- model the effects of the program state.

In principle, one way of modelling the program state is to provide a complete real-time simulation of the environment including all of the times outside of the set of demands of interest as identified by the environment hazard analysis. In this way the correct variation of program state, in keeping with its operational distribution, can be ensured. This approach may be impractical since, although it avoids some of the problems involved with testing the software using the real environment, a time problem remains – for example, nuclear plant protection system software must be tested with environmental events that happen once every thousand years on average (because of the risk presented by such events – see 8.5.3). Therefore, an environment simulation which faithfully reproduces the operational distribution of all environment developments will not perform enough testing for the developments with the greatest harmful consequences since they tend to be very rare. That is, risk analysis may sometimes dictate that absolute environment reproduction is impractical; a numerical example illustrating this effect is developed in 8.5.3. There may be some scope for simulation acceleration, as discussed in Section 8.8, but this may have unexpected consequences, perhaps the most likely being that the software would not be able to cope with significantly speeded up inputs.

The most desirable situation would be if it were possible to reason, for a particular case, that the program state was irrelevant, or of such limited relevance that the important variations could be modelled and enforced by the environment simulation. Such reasoning could perhaps be achieved using formal methods (i.e. formal proof techniques). Unfortunately, with current systems, the way in which program state affects the subsequent ability of the software to perform its functions is highly complex, making this another important area for future research. In future it may be possible to design systems so that this state problem can be more easily dealt with, but this requires further research.

In practice, to overcome the time problem associated with the presence of program state, some assumptions are unavoidable. These will be application dependent. For example, for a protection system, demands might be formed from demands as described earlier in 8.5.1, preceded by various lengths of normal operating conditions, but certainly not the lengths of normal operating conditions which would be expected during the real operation of the plant. Thus the assumption is that large areas of normal operating conditions have no effect on the behaviour of the software during demands.

The Test Result Decision Problem

For each demand, it is necessary to decide whether the software has failed or succeeded. Because of the huge test requirements of SST, this decision will usually have to be made automatically using an 'oracle', which makes decisions 'on the fly' (as tests are conducted). This problem is normally described in terms of a comparison of observed software outputs with those predicted by the software specification for the particular inputs fed to the software. In fact a simpler approach may sometimes be taken. Under some circumstances it may be more convenient, based on notions of controllability, to monitor certain environment attributes rather than program outputs. This would not be appropriate in the case of a protection system where modelling the effects of the intervention of the protection system on its environment is not necessary. However, in the case of a software control system, it is necessary to model the effects of the software on the environment irrespective of the test decision problem. Therefore it might be sensible to monitor variables in the environment to decide whether the software has performed its task satisfactorily.

8.5.2 Statistical Inference

The discussion in this chapter is based on the situation where all test demands are dealt with successfully by the software, although it is possible to modify the analysis to estimate in the presence of demand failures. This is because for safety-related software, it would be unacceptable to allow a known error to remain. On discovery of a failure, the corresponding errors should be corrected, and all testing reperformed. The results reported by Miller et al. regarding their single urn model (SUM) estimator [55] are summarised below.

The distribution for the statistic θ the 'proportion of all demands which fail' (i.e. the probability of failure on demand) is given in equation (1), in which '|x=0' is a reminder that this distribution is only valid for θ when the number of observed demand failures x is zero. t is the number of successful demands executed.

$$f(\theta|x = 0)(1 + t)(1 - \theta)^t, 0 \leq \theta \leq 1 = 0 \quad (1)$$

Accordingly, the expected value of θ (i.e. the best estimate), denoted $\bar{\theta}$, is given in equation (2).

$$\bar{\theta} = \frac{1}{2+t} \quad (2)$$

A measure of confidence in this estimate, the variance denoted σ^2, is given in equation (3).

$$\sigma^2 = \frac{t+2}{(t+2)^2(t+3)} \quad (3)$$

An alternative way to express the confidence is using a confidence interval. In this case, solving equation(4) gives an interval $[0,\theta_c]$ in which we are $(100c)\%$ confident that the true value of θ lies.

$$c = 1 - (1 - \theta_c)^{t+1} \quad (4)$$

A general study of interval estimation is given in most statistics textbooks, for example in [70].

8.5.3 SST for Risk Analysis

The demand based approach allows an alternative approach to that discussed above where probability of failure was estimated for the software as a whole. The alternative is to take standard risk analysis techniques, performed at the system level, down to the software. By 'system level' it is meant that the software is a component of a wider system which surrounds it (and therefore forms its environment). Risk analysis puts a different emphasis on the calculations. Instead of estimating a single failure probability for the software, the emphasis turns to minimising the level of software-associated risk, as justified by the testing. Hazard analysis of the software environment (i.e. the wider system within which the software sits) identifies some hazard-causing events as potentially more hazardous than others. To minimise risk it may be necessary to test more often using the events with the higher severity hazards.

The procedure of risk evaluation is as follows. System level hazard analysis identifies events for which the software intervention is required. In other words, these events are the software environment hazards discussed in 8.5.1 and they constitute demands on the software. Instead of simulating these demands within one large simulation, they are categorised into classes, and a separate simulation experiment is conducted for each class to establish a pfd for demands in that class.

The risk associated with a single specific demand class can be studied, but more often we are interested in a measure of the 'overall risk' associated with using the software i.e. a kind of averaged risk over all requirements. This can be described mathematically as in equation (5), in which the space of demands is partitioned into disjoint demand classes such that each class contains demands of 'the same type', and also within each class the consequences of software failure on any demand of that class are the same. To take the example of a nuclear protection system, a demand class might correspond to a LOCA (loss of coolant accident) event in the plant, with a rate of coolant loss in a given range.

$$\text{risk(per demand)} = \sum_{e \in H} pfd_e \lambda_e \varepsilon_e \qquad (5)$$

In equation 5, e is a demand class; H is the set of demand classes; pfd_e is the software's probability of failure given a demand from demand class e (i.e. a probability of failure on demand for demand class e); λ_e is the probability of the next demand belonging to demand class e; and ε_e is the level of consequences which would occur should the software fail on a demand from class e.

The important consequence of generalising from pfd analysis to risk analysis is that it becomes desirable to test the software in a manner which is not consistent with the operational distribution of the environment. This can be illustrated using a simple example. Take the example where there are 2 demand classes: $e1$ and $e2$. Demands in class $e1$ occur with probability 0.1 and consequence 1000. Demands in class $e2$ occur with probability 0.9 and consequence 10. Testing with the operational distribution of demands and 1000 test demands, there would be approximately 100 tests in class $e1$ and 900 in class $e2$. Therefore $pfd_{e1} = 1/102$ and $pfd_{e2} = 1/902$, according to the SUM (see 8.5.2) and the risk evaluates at approximately 1.0 (risk is measured on the scale 0 to ∞). However, note that with 900 demands in $e1$ and 100 in $e2$, the risk evaluates at approximately 0.2. Thus if the numbers of tests did not have to be in accordance with the operational distribution, testing would usually be able to justify a lower risk value, for the same testing effort.

The above example may give the impression that testing to minimise risk is a matter of convenience only. However, as already stated in 8.5.1 this approach may be unavoidable. Consider the nuclear plant protection system.

It is required to react to some demands which are very unlikely to occur during the lifetime of the plant. Nevertheless testing for these demands must be undertaken because their possible consequences are so severe. Testing must also be performed for the protection system under normal plant operating conditions. However, the required testing for these conditions is low because, despite their frequent occurrence, the consequences of the protection system shutting down a normally functioning plant are low in safety terms (such action may be costly to the plant owners, but does not present much of a threat to safety). That is, testing in accordance with the operational demand distribution is not desired, nor is it possible, since it would take many decades of continuous testing to experience the demands of interest occurring between vast periods of normal plant operation. This is unfortunate since it is only reasonable to use risk-minimisation methods if certain assumptions hold regarding the role of software state. If no such assumptions hold, operational distribution testing can still in principle be used since it is simply necessary to run a single simulation (including all demands) long enough until the level of testing of each event class is high enough. However, as explained earlier, this is unlikely to be a practical solution since, in general, it will require an amount of testing which is not achievable. If, on the other hand, the necessary software state assumptions can be made, the risk-minimisation testing approach is attractive because it fits the quantified safety analysis of software within the standard framework for quantified safety (i.e. risk) analysis.

8.6 The Notion of Confidence in Statistical Software Testing

There are two notions of confidence in SST. Firstly, given a valid statistical model the fact that there is limited test data means that estimates of pfd will only be accurate to some tolerance. The measure of confidence in the estimate has been discussed earlier in 8.5.2. It is the usual concept of confidence in statistics.

Secondly, there is confidence in the validity of the statistical models themselves. For SST, this type of confidence is not well understood. There are two approaches which can be used to establish confidence of this kind:
- the study of the assumptions on which the models are based
- the empirical validation of models.

Empirical validation of statistical models poses an interesting problem where those models are estimating the pfd for a safety-related system. The rarity of failure events means that pfd cannot be directly measured at the levels required using operational systems. Consequently, it would be fair to say that there is no very high integrity software of any complexity whose pfd is known, we have estimates only.

8.7 Criticisms of Statistical Software Testing

A common criticism is that statistical testing cannot be applied to software. The argument is usually based on a comparison with statistical testing of hardware components for faults of a particular nature i.e. wear-out effects, or defects in raw materials. Faults in software are not of this nature, they are design faults. Thus, it is sometimes argued, software faults are either present or not and there are no elements of uncertainty present for probability theory to model. It is in the last step in that this argument falls down. The uncertainty concerns how often any faults will be exposed during normal usage. Indeed, if we take a new untested hardware component which is to be used in a known environment, it could be argued that it will already 'contain' (or not, as the case may be) its faults whether material defect faults or even wear-out faults. Thus this common criticism of statistical software testing is not well founded; the premises are true but the conclusion, namely that statistical software testing is impossible, is not warranted. However, this is not to say that the fact that software faults are design rather than wear-out faults has no significance.

In fact there is another criticism which is well posed, but is an open question. The crux is the extent to which the test results are evidence towards the behaviour of the untested inputs. In this respect statistical testing of software and hardware is very different. We know from experience that samples of hardware components do give us an indication of the failure behaviour of a new component. The reason this predictive inference works is because the new component is made the same way and is subjected to the same environment, and we have models, based on the notion of 'wear-out', of component failures which suggest that it is these features of components which are important. That is, there is very good reason to believe that the failure behaviour of similar components in a similar environment is relevant evidence for the prediction of new component failure behaviour. However, in the software case the question is whether one test set is relevant evidence for inference about the likely failure behaviour in other parts of the software input space. All statistical approaches, including the current ones, must assume something about the relevance of tests performed to those not performed, and these assumptions require validation. Because software now commonly forms a part of many safety-related systems, and because SST is the only technique to offer a clear objective quantified measure of integrity based on the sound foundations of probability theory, further work is needed on this particular topic which is the foundation of any form of SST. It is possible that new statistical models could achieve higher reliability estimates from a given set of successful tests.

In summary, the most common criticism of statistical software testing is invalid, and really we should be concerned with another question: are the probability models on which the statistics are based true to life? Statistical software testing is possible in principle, but it relies on tests being relevant to each other, and the existence of models describing that relevance. There is no work in the literature which convincingly argues against the possibility of SST. Some interesting introductory ideas to the subject are given in [71].

8.8 The Future of Statistical Software Testing

In the field of software testing, SST stands out because it provides results which can be used directly in quantified safety arguments, and in addition it is based on the sound mathematical foundations of probability theory. For these reasons SST is currently an important technique, and is potentially the crucial testing technique for safety-related software. For software whose required level of justified reliability is not too high, the current state of SST already provides a practical means of conducting quantified safety analysis for systems containing software components.

To carry out SST, is to perform, firstly a statistical testing experiment, and secondly statistical inference on the experimental results. Proposed appropriate statistical inference procedures are already in existence, although there are ongoing developments in this field ([69] and [72]). The main effort involved in SST is in the setting up of the testing experiment, where simulation of an operational input distribution for the software is necessary. Practical techniques to achieve such a simulation are described ([73], [74] and [76]) but a related task is the connection of such a simulation to the software to be tested. This remains a manual task with little tool support. Certainly current test harness tools are not sufficient, relying as they do on lists of test cases, or, at best, simple test scripts controlling the presentation of many lists of tests to the software. Test by environmental simulation, in general, requires coincident evolution of both the software and its environment where the actions of either entity is dependent on the other.

Although SST can already be applied with some confidence in certain circumstances, there are many issues which remain unclear. For example, it is possible with current SST techniques for large areas of code to be left unexercised. If this were to happen, it would be sensible to either continue to increase the number of randomly chosen tests until all code is covered, or choose specific tests to achieve coverage. Considerations such as these suggest that the statistical models are to some extent incomplete – that they possibly need to include some white box testing considerations within them. Partition testing is intuitively felt to be sensible and effective by most software engineers. If it is effective we should be able to build statistical models to account for the improved testing that results. Certainly such a theory does not exist. In fact when Miller builds a statistical model for partition testing the estimated pfd is higher, given the same test results, than the estimate for non-partition testing! The reason for this is probably that the modelling in the partition case is far too conservative. Other related work is also non-intuitive: Duran & Ntafos, [44] find that partition testing is no more effective at finding faults than random choice of tests. This could be because of some radical assumptions made to simplify the analysis. Any new statistical models, such as models of improved testing via partitioning, must follow from a better understanding of software failure mechanisms as discussed in 8.7.

Finally, the subject of the limitations of SST is currently receiving much attention. Littlewood has shown that assuming a particular statistical model of the failure process there is a probability of 1/2 that a program will not fail in a time t, given that it has already been tested for a time t and no failures have occurred [56]. This clearly rules out the possibility of using those SST models for demonstrating ultra-reliability. However, it should be noted that, as always, this is a result based on a particular statistical model. Furthermore, justified reliability figures depend on the rate at which tests can be conducted so that, where tests can be performed more quickly than they would occur during real software usage, high reliability justification is possible. For example, according to the Miller et al. model, if the duration of a single demand was ten seconds on average, two months of testing would result in a best estimate (expected value) probability of failure on demand of 3.7×10^{-6}, and a 99% confidence that the pfd was in the interval $[0, 8.6 \times 10^{-6}]$. It might be feasible, for example, to perform simulation of a day's operation (considered as a single demand), within ten seconds. However, to achieve a justified pfd of 10^{-9} at confidence 99% in two months would require 1000 tests per second, so that ultra-high reliability remains beyond reach in practical terms. Where demands are lengthy, the Miller et al. approach implies that speeded-up simulated environments are very attractive, providing it could be argued that the increased speed of execution did not effect the test results in any way. Test acceleration is an area requiring further investigation. It may be premature to state that SST is limited to justification of, say, software reliabilities of 10^{-4} or less. The issue is dependent on a better understanding of statistical modelling of software failure. Meanwhile it is important to obtain accurate estimates of integrity at levels which are feasible, and many practical problems are in this category e.g. justification of certain software protection systems for nuclear power plants.

The following issues have been identified to be of crucial importance because, successfully addressed, they would make SST a practical solution to the justification of safety-critical systems under a much wider range of circumstances than is currently the case:

- analysis of the role of software state. Either methods of building software so that its internal state does not result in different software behaviour to the same demands; or methods of modelling the progress of program state and its effect on the statistical failure behaviour of software
- empirical validation of statistical models of software failure
- new statistical models to account for the relevance of tests to other areas of the software input space, i.e. new foundations of SST based on physical arguments concerning the features of software. These models would include white box considerations in SST (currently a black box exercise) perhaps to guide a partition testing strategy
- conditions under which test acceleration is valid.

8.9 Guidance

- Recreate (simulate) the operational profile or demand space of the software.
- It may be necessary to test using inputs beyond those of direct interest, depending on the influence of software state.
- If an environment simulator is used for testing software then an automated oracle will almost certainly be needed.
- Choose an appropriate single-version testing model to use the available test data.
- Ensure the assumptions on which the model is based apply.
- A statistical inference should be conducted from observations of the software's performance in the simulated environment to produce an estimate of the probability of failure.

Chapter 9
Empirical Quantifiable Measures of Testing

In this chapter, two empirical quantifiable techniques for assessing software integrity are discussed; test cost assessment and test regime assessment. The first assessment method provides guidance on the estimate of test costs whilst the latter offers a systematic method of building up an argument for the contribution which testing can make to the safety case.

9.1 Introduction

Test cost assessment aims to provide a framework in which developers may identify the test costs consequent on adopting particular design features. The effects of a system's design features on its testability with respect to its proposed Test regime are discussed and a system 'scoring' procedure is proposed. The resulting metric is not aimed at measuring a system's quality, it is intended to provide guidance in the estimation of test expenditure.

Test regime assessment offers a systematic method of building up an argument for the contribution that testing can make towards a system safety case.

9.2 Test Cost Assessment

The level of safety integrity that can be claimed for a system is influenced by the tests performed and *'successfully'* completed. Thus it must be possible to perform all the tests specified for a required level of integrity. To perform these specified tests the system must be designed appropriately. Using a definition of testability of *'The ability to perform all the specified tests'* clearly does not help in forming a scale of testability as all systems should be designed to meet this requirement. To derive a scale it is necessary to have some means of discrimination. It is suggested that this be achieved by using the ratio of design expenditure to design plus testing expenditure.

$$Test\ Cost\ Factor(TCF) = \frac{Test\ Expenditure\ (TE)}{Test\ Expenditure\ (TE) + Design\ Expenditure(DE)}$$

TCF will have a theoretical range of 1 to 0. However a value of 0 will never be achieved as test effort will never be zero. Brooks, in his book *The Mythical Man Month* [77] suggests that testing activities should take up to half the total project effort whilst Boehm [24] gives a more complex statement of work breakdown, but as an indication, for a 'very large embedded mode development', it suggests around 22% of total effort be expended on test related effort. Boehm suggests that the percentage will vary with the required system reliability, due to an increase in effort during integration and test relative to other activities, by a factor of up to 1.75. The figures given by Boehm are based on a range of projects and are not specifically based on real-time safety-related systems. The TCF may be expected to vary with the required Safety Integrity Level (SIL), the higher the SIL the lower the value of TCF, caused by more onerous testing requirements. The relationship between TCF and system size or complexity is less clear because, as system complexity increases both design and test effort will increase thus having a limited effect on the TCF. It might be that *Test Expenditure* increases at a much higher rate than *Design Expenditure* for given increases in system complexity. However, although *Test Expenditure* is potentially limitless from the technical point of view, there will be economic limits. A project in which testing costs are three times those of design would be rare. Thus TCF should have a practical range of 0.75 to 0.25.

The TCF should be calculated for a number of projects to create empirical models that determine reasonable values for each recognised integrity level. The *Test Cost Factor* does not measure a product's quality, it measures an organisation's adeptness in designing for testability. Purchasers should monitor a developer's track record of TCF for systems certified to appropriate levels of integrity to judge whether the developer can design for and execute a required *Test Regime* efficiently.

To calculate and use the TCF method, it will be necessary to:

- record costs against design or test activity
- record costs of equipment used for production or test.

To obtain a relationship between a system's design features and its testability two further steps are necessary:

- record the test expenditure for a variety of projects identifying common and diverse design features
- establish a correlation between the *Test Expenditure* and the design feature(s).

In order to assess whether design features have a significant effect on system testability as defined here, it is suggested that a list of primary attributes of systems be drawn up and the TCF be grouped by primary attributes and reported for each system assessed. For example, the I/O count could be recorded for each system and the TCF could be reported for systems with I/O counts in identified ranges.

There are a number of PES types and what can be regarded as a primary attribute of a system may vary from type to type. A discussion of typical design features of PESs may be found in Chapter 3. From these discussions a number of questions which can be used to form a testability check list in conjunction with those of *Guidelines for Assuring Testability* [78] are raised. These testability questions should be considered when estimating test expenditure. If as expected, a significant link between design features and test expenditure is shown to exist, then such information can be used to support design decisions on future projects. This approach is similar to those proposed for assessment of reliability of software based on process and product properties in BSI DD 198, 1991 [79].

The questions fall into two categories: those with a quantitative answer, e.g. 'How many modules?', and those for which a descriptive answer is required e.g. 'What equipment is required to exercise the I/O points?'. For the quantitative answers the effect on TE is obvious, the more components there are in a system the greater test expenditure will be. For descriptive answers, the contribution to test expenditure is subjective. For example if a certain item of test equipment is required, then the cost, availability and user familiarity all have a bearing on the test expenditure.

9.2.1 Calculation of TCF during the Development Lifecycle

There are particular stages during development lifecycle when it is useful and practical to calculate the TCF.

The value of TCF should be calculated from tender estimates to give an indication to both supplier and purchaser that adequate provision has been made for testing. After the tendering phase the TCF can be re-estimated as design decisions are finalised supported by data (experience) previously obtained.

It would only be practical to calculate TCF for a system/subsystem/module during development if the matching design and test activities are complete, otherwise the ratio may become distorted as expenditure in one area outstrips expenditure in the other; it is not intended as an on-line measure. Estimates of TCF can be made for each completed stage of development. However due to the large variation in test expenditure that could be reasonably expected between Requirements and Integration Phases for example, its value as an indicator may be limited. Thus it is thought inappropriate to calculate TCF during development. Bearing in mind that testing of higher integrity level safety-related systems could potentially cost more than its design and manufacture, the importance of ensuring that adequate provisions for support of testing have been made cannot be over emphasised. Furthermore if independent testing is required it is likely that a separate testing contract will be required. In such a case, the assessment activities described here would form part of the tendering activity for such a contract.

The calculation of TCF on completion of development is regarded as straightforward and desirable as a check against the initial estimates.

Thus the recommendation is to calculate TCF at:
- tender completion, to provide evidence of adequate test provision

- completion of top level design phase, to ensure that the design decisions made are broadly in line with the tender commitments
- contract completion, to provide feedback to the estimating procedures employed at the two previous calculation points.

In the following paragraphs the calculation of TCF at each of these points in development is considered in light of the likely knowledge of cost drivers (design features) discussed earlier in Chapter 3. The intention is to suggest dependencies between cost drivers introduced at each phase of development and *Test Expenditure*. Exact relationships are not sought, for at best, these relationships will be specific to an organisation. It is likely that organisations would not have sufficient data which would allow the exact relationship to be determined, hence the suggestion that a data gathering process is instigated to 'calibrate' the measure within a particular organisation.

9.2.2 Calculation of TCF at Tender Completion

At the tender stage an assessment of the standards pertaining to the system under development must be made; as these standards greatly influence all matters of system development. They affect the physical (hardware) build of a system, the logical (software) build of the system and the methods to be employed in developing the system. To this end the *test regime assessment* model (which is discussed later in 9.3) may be used to identify the tests necessary to satisfy certification authorities. The Test Regime will clearly have great influence on the way a system is implemented so that all the test criteria can be met.

In estimating TE the following steps are suggested:

- identify the required integrity level of the system.
- identify required testing methods.

Options may be available. Details of how to develop an argument for a test regime commensurate with a required SIL are given in 9.3. An assessment of the best test strategy may be made on the basis of the outline system design.

- identify in-house expertise in the test methods required
- estimate the required system test expenditure on the basis of the previous steps.

This last step is far from trivial although it is carried out regularly in tendering activity. If the TCF value at this stage does not match a level thought appropriate for the SIL then steps should be taken to adjust it; such steps could include:

- making a case for lower SILs for the PES or parts thereof
- choosing alternative test methods
- changing the system architecture or adopting alternative technology
- training or recruitment of staff.

9.2.3 Calculation of TCF at Completion of Top Level Design

On completion of the top level design of a system, a better understanding of the components of the system will be available and so test expenditure can be considered in greater detail. At this stage there may still be room for manoeuvre as far as the design and the testing regime is concerned. Beyond this stage the overall system architecture and the interfaces available to support testing will be fixed. Omission of adequate interfaces may affect the ability to meet the Test Programme within budget. It is suggested that a major review of the system design is carried out if the TCF measure differs significantly from the tender value, which would be based on typical values for similar systems.

A discussion of cost drivers regarded as significant in the estimation of test expenditure for each of the subsequent development phases is given in the following paragraphs.

Design Phase Contribution to TCF

At this stage of development 'testing' will be carried out by review of the design. In some cases animation of the specification may be employed
Test Expenditure dependencies include:

- system I/O capacity
- number of subsystems
- number of Control Decisions.

These features affect not only the design complexity of the deliverable system but also the design complexity (and classification) of any simulator required to test the system.

There are qualitative aspects of design features which can also be considered at the design phase, i.e.:

- the nature of any simulator required to perform tests on the system
- the hardware interfaces intended for the PES, which will affect the expenditure required to build a simulator.

Module Design Phase Contribution to TCF

Review techniques will again be used at this stage of development. Prototyping may also be a valid method of testing at this phase.
Test Expenditure dependencies for this phase of development include:

- number of functions supplied by the module
- number of events monitored/used by the module
- complexity of functions supplied by the module.

The qualitative aspects of the design of a module which contribute to test expenditure include:

- The requirement for a formal specification of a module. Checking of such specifications requires very specialised skills.

- The use of graphical methods to express functionality which tends to reduce the time taken to assimilate design decisions prior to design review.

Module Test Phase Contribution to TCF

Testing of modules may be carried out statically by analysis and review of code or dynamically by execution of the module in a host or possibly in its target system.

Cost drivers relevant to this phase of development include:

- number of modules
- module complexity
- coverage criteria.

These 'cost drivers' are measurable using objective methods. Further dependencies which are more difficult to assess include the choice of language and tool availability (simulators for example), the interdependence of modules and the traceability of functions to modules.

Integration Phase Contribution to TCF

At this stage of development, testing will be focused on dynamic execution of the system and its subsystems. In particular, timing of events and recording of responses should be carefully considered. It is important that the TCF is calculated after top level design, and so attempting to foresee timing difficulties at integration testing time may be very difficult. The use of simulators/animators at the top level design stage may help considerably in assessing testing problems related to timing.

It is suggested that cost drivers include:

- number of modules
- number of control loops
- coverage criteria (black box)
- number of displays
- system I/O capacity
- observation frequency
- number of control decisions.

Other cost drivers which have an effect on test expenditure at Integration Phases are: the availability of automatic test equipment, the PES architecture, and the mode of operation of the PES.

Validation Phase Contribution to TCF

At the validation phase significant dynamic testing occurs and there is considerable expenditure on quality assurance, checking that all phases of development have been carried out correctly and that supporting documentation exists. The influences on test expenditure required at the validation phase include:

- integrity level
- number of documents detailing system design and implementation.

Less quantifiable aspects which contribute to the validation phase are: the use of formal/structured methods, the software structure, the PES architecture.

9.2.4 Calculation of TCF at Contract Completion

Calculation of TCF at contract completion will depend on the availability of adequate records of expenditure on design or test. Easy access to such information will depend very much on the extent of the management information system employed. Provided that expenditure is logged against a project accurately and with sufficient discrimination, the calculation of TCF at this stage can be regarded as straightforward.

9.3 Test Regime Assessment

The two approaches to safety assessment, process and product assessment, can be used in conjunction to improve confidence that a particular system is sufficiently dependable. It is suggested that testing, which has the advantage of being able to give tangible evidence of the behaviour of the whole system (hardware, application software and system software, including their interactions, restrictions and timing), can contribute to both arguments.

While product methods (such as reliability growth modelling and random input testing as described in Chapter 8) might be effectively used for systems with the lower levels of safety integrity, they are likely to be impractical for systems with higher levels of safety integrity (SIL 3 or 4). Note that this does not preclude their use to give confidence that a system should at least reach these lower integrity levels. Therefore, if it is not possible to quantitatively measure software integrity as a reliability figure, the aim must be to gain confidence that the required level of integrity should have been met by qualitative means. However, the problem then remains of how qualitative arguments can be reconciled with a quantitative requirement for probability of failure on demand or frequency of unsafe failure as a part of numerical safety cases.

The current approach of the certification authorities is to use engineering judgement to assess the process which produced the software system and then to assume that the integrity of the product is related to the quality of the process. This assumption is based on the three aspects inherent in a high quality process which aims firstly to avoid the introduction of faults, secondly to detect them should they occur, and thirdly to tolerate any faults which might never the less be built into the system. It is evident that testing, using both dynamic and static techniques, is the major means by which faults are detected, and so it is argued that significant effort should be directed at providing some measure of the quality of the second aspect – fault detection. A process could then be considered to be of sufficient quality for a system at a particular level of integrity if each of the three aspects are shown to have been sufficiently well addressed for that level.

For error detection, it is suggested that the best approach is for the method to reflect the system testing performed and thus to present a qualitative assessment of the testing process taking evidence from a combination of many testing methods (functional, structural, static). This evidence must be used in an auditable and repeatable way if it is to be credible as contributing to a safety case.

9.3.1 Test Regime Assessment Model

The proposed method is based on the *partial beta factor model* agreed by the safety authorities for the quantification of Common Mode Failure (CMF) of high integrity systems in a number of defence applications which has been derived from SRD's work on dependent failures. The numerical assessment of CMF is a very similar problem to that of quantification of software failures as a purely qualitative examination of a system must be used to derive a failure probability rate based on engineering judgement and experience. In addition, it is noted that both CMF and software failures are due to systematic errors rather than random errors and that the sources of these errors are very often similar or the same. It is also noted that the two problems have been addressed in much the same way and that the method of assessing CMF seems to be being resolved in favour of a partial beta factor method. SRD describe the Partial Beta Factor method as a structured way of estimating a system-specific value of Beta, based on the quality of system design and extent of defences against dependent failure, which has been used in the UK for reliability assessment of both nuclear and non-nuclear plant. To apply this to the software problem, it would seem that a structured way of estimating a system-specific value for software reliability based on the quality of system design and the extent of defences (fault avoidance, detection and tolerance) against software failure, while perhaps not ideal as it is a measure of the process and not the product, might be considered to a worthwhile advance on current practice.

The method used is subjective, but it is considered that an appropriate weighting of the factors involved can ensure that it is conservative and so will not result in over-optimistic evaluations of software systems. Note that 'conservative' is used relative to the level of claim that certification authorities might consider appropriate rather than in terms of the achieved reliability of the software system. The method attempts to limit the amount of choice for individual analysts by putting as much as possible of the subjective element into the model where it can be checked and modified to reflect experience.

The model described below is appropriate for safety-related systems that operate on demand (e.g. protection system). A similar model can be constructed for continuous operation systems.

The first step in deciding the form of model to be used is to decide the range of values of probability of failure on demand (pfd) which can be considered applicable to software systems. The minimum level (i.e. best claim) of pfd that can be reasonably justified for a system developed using present techniques has been set at 1×10^{-5} as this is the extreme limit of safety integrity level 4 (SIL 4) given in the IEC draft standard [6]. It may be that another level would achieve a greater consensus, but the figure has been chosen for consistency with the IEC standard and others. The model factors have thus been chosen in ensure that the resultant claim can not be significantly better than 1×10^{-5} pfd (and obviously not larger than 1).

The model decomposes the contribution of testing into a number of primary factors related to fault detection/testing:

Functional Testing Black box tests designed to ensure that the product reflects the requirements and design. This approach represents the primary means of validation.

Structural Testing White box tests designed to ensure that the product reflects the design and thus contributes to verification at this lifecycle stage.

Static Analysis These checks ensure that the product reflects the design and that the design reflects the requirements. Thus these techniques are capable of contributing to verification at each lifecycle stage.

Environment This is a weighting to reflect the appropriateness of the testing environment particularly with regard to addressing hazards. The better the environment the more likely that real input distributions will be reflected and thus that errors likely to occur in practice will be detected – i.e. those errors likely to have a significant effect on the system and on its performance of its safety functions.

Testability This is a weighting to reflect the testability of the design. The more testable the design is, the more likely that the testing will detect errors.

The product of these primary factors (functional, structural, static, environment and testability – F_1 to F_5 respectively where $1 \geq F_i > 0$) represents the claimed software integrity I_T (probability of failure) based on the testing process.

$$I_T = \prod_{i=1}^{5} F_i \qquad (6)$$

The five primary factors (F_1 to F_5) are further broken down into their secondary factors (f) which have individual values apportioned based primarily on the weightings from the tables given in the draft IEC standard [6], but bearing in mind I_T is defined within the range $1.0 \geq I_T \geq 1 \times 10^{-5}$. The analyst determines the software reliability by deciding (and justifying each decision) to what extent (as a percentage) each of the individual test criteria have been satisfied. Each of the test criteria has a minimum claimable factor (f_{min}) associated with it, and the actual factor to be used (f) is determined by reducing this claim in line with the justified compliance factor percentage (P).

$$f = 1 - (1 - f_{min}) \times \frac{P}{100} \qquad (7)$$

Thus each secondary factor will be in the range $1.0 \geq f \geq f_{min}$. The overall factor for each of the five primary testing factors is then determined by taking the product of the individual test criteria factors in a similar way as for software integrity (i.e. equation 6). Note that each of the minimum factors are in effect weightings for the relative benefits of each testing approach, and that these minimum factors have been scaled so that their product is of the order of 1×10^{-5} because this is believed to be the best integrity achievable for demand based systems using current techniques. These minimum factors allow the more significant amongst them to be identified so that effort can be concentrated on these whilst the less significant factors can be given a less complete but none the less auditable treatment.

In general, it is suggested that a discrete scale for the values of P be employed unless they can be based on some objective measure. For example:

100%	Comprehensive
75%	Mostly
50%	Fair
25%	Partly
0%	Not addressed

The choice of P should include consideration of:
- the proportion of software subject to the technique
- the lifecycle stages at which the technique was used against those when it could have been used but was not
- the expertise and independence of the personnel involved
- the depth to which the technique was used (e.g. a review might involve one person for an hour or five people for a day)
- the level and quality of tool support
- the complexity of the software subject to the technique.

If it is considered that a particular technique is not applicable to the system under consideration then it is suggested that an assessment is made as to why this is the case and a value of P assigned accordingly. For example, boundary value analysis could be considered to be not applicable for a purely digital system. In this case 100% compliance could be claimed as this type of error cannot occur and so might be argued to be 100% detected. Alternatively, information hiding and module size limit could be not applicable for ladder logic code. However, in this case these techniques are advocated to reduce the impact of small changes and to aid analysis respectively. Both of these benefits are lost with ladder logic and so 0% compliance is appropriate.

It is suggested that the issue of changes in design after tests, might be incorporated by reducing these values of P by a number of steps in line with the extent of the modification to reflect the reduced level of confidence in any tests performed prior to the modification. It is current practice to require a full retest of the whole system following modifications and while this is undoubtedly the safest course and would still be mandated for any changed module, it may be excessive to retest the whole system at lower integrity levels. For example, if a 75% compliance with a functional testing criteria had been claimed but one of the modules was subsequently changed when a fault was found, then the changed module would be retested and the level of compliance might be reduced to 50% to reflect the possibility that the modification, whilst not affecting the individual module, might have some unexpected effect on the overall system. All such affected factors would be altered and the claimable integrity recalculated. If the level was still satisfactory then no further action need be taken. However, if the level is not satisfactory, then some retesting of the whole system would be necessary. This suggested approach would ensure that retesting would still be necessary at the higher levels as the claimable integrity would be significantly reduced but may allow a lower, but still acceptable, claim to be made at the low integrity levels (if there was originally some margin after testing) without incurring the extensive costs involved in retesting the whole system.

An argument can be made against the use of the product of the various factors because the tests may be revealing the same faults. This problem can be expected to be exaggerated for similar methods. It is considered that the optimum approach to this problem would be to assess the individual testing techniques' absolute contributions to error detection and then to combine the factors using some overlap factor appropriate to the degree of similarity between the methods. However, this would be difficult due to the large number of combinations of methods, all of which would have a different degree of overlap, and also there is not believed to have been any theoretical work dealing with this problem to any degree.

The software integrity as predicted using the model should then be confirmed by random testing (from the distribution of test cases where correct operation is important) to show that it is no worse than that claimed. Similarly, reliability growth models could be used to show that the integrity is no worse than that claimed. Where problems exist with these models, or with the amount of random testing needed for demonstrating very low failure rates, then they might be used to give confidence in some lower limit to the software integrity. Note that it is considered that at least one of these techniques should be adopted regardless of the integrity level. This is because the volume of tests required increases as the integrity level increases, and so for low integrity levels the costs of these approaches should not be prohibitive. Indeed some of the standards recommend the use of the techniques.

9.3.2 Assignment of Factor Values

The first three primary factors of the model (functional testing, structural testing and static analysis) have had minimum factors derived from the scores for the tables in the IEC draft standard and bearing in mind those overall scaling considerations (i.e. no claim better than 1×10^{-5}) discussed earlier. Note that the minimum factor for each table is the product of the minimum factors of the testing techniques contained within that table and represents a minimum factor for the particular approach to testing.

Functional Testing

Table 9-1 : Functional Testing (Minimum Factor = 0.044)

Technique	Minimum Factor
Test case execution from boundary value analysis	0.7
Test case execution from error guessing	0.8
Test case execution from error seeding	0.9
Equivalence classes and input partition testing	0.7
Test case execution from cause consequence diagrams	0.9
Prototyping/animation	0.9
Process simulation	0.8
Avalanche/stress testing	0.8
Response timings and memory constraints	0.6
Performance requirements	0.4

Structural Testing

Table 9-2 : Structural Testing (Minimum Factor = 0.4)

Technique	Minimum Factor
Structure based testing	0.4

It is suggested that the percentage conformance claimed for structural testing should be based on the coverage achieved. For example, a set of values might be:

No predefined level of coverage	0%
Statement coverage	20%
Decision coverage	40%
Condition coverage	50%
Decision/condition coverage	60%
LCSAJ	60%
Multiple-condition coverage	80%
Basis path testing	80%
Exhaustive all path testing	100%

It is likely that 'interface testing' should also be included under structural testing (white box) techniques and should have a weighting of perhaps 0.7.

Static Analysis

Table 9-3 : Static Analysis (Minimum Factor = 0.05)

Technique	Minimum Factor
Boundary Value Analysis	0.7
Checklists	0.8
Control flow analysis	0.7
Data flow analysis	0.7
Error guessing	0.8
Fagan inspections	0.9
Sneak circuit analysis	0.9
Symbolic execution	0.7
Walkthroughs/design reviews	0.6

It is considered that there is a strong case for 'information flow analysis' to be added to the static techniques. This approach is recommended in most standards and so could be included here, perhaps being given the same weighting as the 'control flow' and 'data flow' analysis methods. it is also possible that 'response timings' and 'memory constraints' might be better allocated as a static method rather than as functional testing.

The values assigned to the final two primary factors of the model (Environment and Testability) are more tentative. However an initial assessment has been made and the results listed below.

Environment

Table 9-4 : Environment (Minimum Factor = 0.10)

Technique/Measures	Minimum Factors
Level of Testing Independence	0.6
Personnel Training & Experience	0.8
Quality of Simulation	0.5
Consideration of Hazards	0.6
Quality of Tool Support	0.9
Module Test Integration Method	0.8

The idea behind this table is that some assessment should be made of the testing environment as both the efficiency of the test personnel and the closeness of the environment to the plant environment will have a major impact on the success or failure of the fault detection techniques used in the previous tables. These should be addressed when assigning the percentages which reflect the degree of compliance with the individual techniques, but a general assessment is also thought to be relevant.

The two areas in the above table which are perhaps of most interest are the 'quality of simulation' and 'consideration of hazards'. The first of these relates to the test environment as a simulation of actual plant conditions. An ideal environment might be the full software running in the final target hardware on the actual plant in full operation which however is obviously unachievable in the vast majority of cases. A poor environment would test small sections of the software in an emulation of the hardware and system software with test signals being injected without regard to real plant states. This factor thus reflects the likelihood of detecting faults which will affect the final system. It is also important that the simulation reflects the expected real input distributions in the test cases so that rather than just detecting faults, the testing tends towards predicting reliability by removing those faults which may be likely to occur in practice.

The second area of interest, 'consideration of hazards', gives benefit for the degree of traceability between the identified plant hazards and the system test cases. It must include some allowance for the means by which the hazards have been determined with a view to quantifying the completeness of the hazard analysis. To some extent, protection against plant hazards will have been identified within the functional specification and so the testing will be overlapping with the benefit already accrued in the functional testing section. However, this more explicit check against the plant hazards identified by the hazard analysis performed with small perturbations of the input space may be a worthwhile step as it checks the safety requirements in the specification. In other words, an accurate simulation of the plant environment, together with test cases based on a hazard analysis, could contribute to the verification of both the system's execution of its plant safety functions, and also the correctness and completeness of those safety actions in terms of overall mitigation of the plant hazards that the safety system was provided to protect against. It should in any case be extended as far as possible to cover hazards created by the system itself, thus giving a useful tie-in to the system safety case. It is considered that these approaches are sufficiently valuable to be assigned significant weightings as reflected by their relatively small minimum factors.

Testability

Table 9-5 : Testability (Minimum Factor = 0.075)

Measure	Minimum Factor
No dynamic objects or variables	0.8
Limited use of interrupts	0.8
Limited use of pointers and recursion	0.8
No unconditional jumps	0.8
Module size limit	0.7
Information hiding/encapsulation	0.7
Parameter number limit	0.9
One entry/one exit point in subroutines and functions	0.8
Synchronous I/O	0.8
System I/O capacity	0.8
Observation frequency	0.9
Number of displays	0.9

There may be a case for also including 'metrics collection' and/or 'application complexity' in the above table.

System Software

The method proposed above can deal adequately with application software but presents problems for how system software might be approached. System software is primarily considered to mean the operating system. However, it is also possible to extend the definition to include such things as compilers, linkers, loaders, translators and interpreters. The difficulties in assessing system software arise because there is generally a lack of knowledge of the structure of this software which is often obtained from third parties, and so structural testing, static analysis and even testability cannot be dealt with as described above unless some sort of certification scheme of system software was adopted. However, it is not true that these elements of the application software testing provide no degree of confidence about the system software. Structural testing includes demonstrating that the application software together with the system software performs as per the design. Static analysis, which shows that the application code is relatively well structured and simple, suggests that the system software is less likely to be overburdened or challenged by unexpected tasks. Testable code would also enable the same conclusions to be reached.

It can be argued that the best test for the system software (and for the translation of the application software) for a particular system is to test the application software (in its object code form) functionally in as near the final target environment as possible. Thus the system software will be executed in a similar manner as when it is in operation and any problems should be detected. However, as for application software, it is suggested that this is not sufficient for safety-related applications and that further confidence is required.

Therefore, it is suggested that where the system software cannot be covered as described in the model, then an alternative approach of relying on the tried and trusted nature of the product must be adopted. If previous usage is to replace testing as a safety argument then it must constitute an extensive track record in a similar environment, performing a similar task, and with very few modifications having been made to the software throughout its lifetime. This opinion is based on statements made in the past by certifying authorities.

Table 9-6 proposes that a little more than two orders of magnitude are based on the track record and that a reduced factor based on the application software testing which has been performed in conjunction with the system software should also be included (perhaps the square root of the application software claim giving a minimum factor of 3×10^{-3} might be appropriate). However, if some testing has been performed using an emulation of the system software, then it is considered that no (or at most a very much reduced) claim should be made on this testing for the system software. The safety claim for system software would then be calculated as the product of this allowance for the application software testing and the factor obtained from the use of the Track Record table (9.6) with each element being given roughly equal weighting.

System Software Claim =
(Application Software Testing Claim)$^{1/2}$ × Track Record Factor

The use of this track record approach will mean that the minimum claim that can be made for system software is approximately double (i.e. worse) that which could be made for application software. It may be that this claim is still too optimistic and should be limited to 1×10^{-4} or higher where arguments are based on track record rather than on a comprehensive approach to testing. Note that to achieve the same minimum level of claim the system software would have to be thoroughly tested in the same way as the application software.

Table 9-6 : System Software Track Record (Minimum Factor = 0.0051)

Measure	Factor
Experience	0.2
Environment	0.4
Function	0.4
Modifications	0.4
Problems	0.4

'Experience' reflects the number of times and the duration that the system software has been used before (i.e. this criteria gives an indication of the volume of 'test' data).

'Environment' signifies the degree of correspondence between the predicted use of the environment and that which the software has encountered in the past (i.e. how relevant the 'test' data is to the application under consideration).

Similarly, 'Function' reflects the correspondence between the task that the application software is to perform and those functions performed in the track record.

Note that if extensive experience with the software is claimed but that only a proportion of the previous matches the current environment and function, then either the experience claim or the environment/function claims should be reduced. For example, some system software might have been used many times before in aircraft and once for a nuclear protection system. If it was to be used again for another nuclear protection system then the claim should either reflect the extensive experience but with a different environment and function or alternatively very limited experience (i.e. one usage) but with a very similar environment and function.

The 'Modifications' factor is included to enable benefit to be claimed for system software that has an extensive track record but which has undergone small changes throughout its lifetime to correct bugs and/or introduce new facilities. The greater the number of minor modifications over the software lifetime then the smaller the claim. Note that major changes should be regarded as restarting the track record from scratch.

Finally, 'Problems' is a category included to reflect the general experience with the software throughout its track record on the basis that past problems may be indicative of software which is not robust even though the previous problems have been fixed (errors tend to come in clusters). This category might also be used to reflect the attitude of the system software producers to problem reporting and corrections, e.g. are there formal procedures and are customers automatically kept up to date with other users experiences so that the impact of known bugs on safety-related applications can be assessed.

It can be seen that all of these factors have been rated equally as no basis for any other scoring is available. The only exception to this, the actual extent of the track record data, 'experience' seems to be the most valuable evidence and so has been tentatively weighted more heavily than the other factors.

It should be noted that the values and method suggested for system software are particularly speculative and should be considered for each application. The track record approach has advantages in that it can account for existing systems, while new software must either be appropriately tested for the integrity level at which it is to be used or it must have an extensive track record in non-safety-related applications (note that non-safety applications would not score well on the Function entry in the table and thus would probably not be able to be used for the most safety-critical applications until its track record for safety systems had been established).

9.4 Discussion of Test Regime Assessment Model

In summary the model suggests minimum possible factors as follows, although of course the table entries and their weightings should be viewed as speculative:

Functional testing	0.044
Structural testing	0.4
Static analysis	0.075
Environment	0.10
Testability	0.075

These factors give a minimum achievable claim of 9.9×10^{-6}.

It is considered that different sets of minimum factors might be applicable to different types of systems although the differences may not be very marked. An obvious example is that structure based testing is not normally used for ladder logic, although perhaps there are ways in which it might be applied slightly differently (i.e. all external coils [outputs], all coils [statements], all special functions, all inputs [decisions]).

There is a question of how new testing methods could be incorporated into the model given that the factors have been carefully scaled to limit the greatest possible claim that can be made using the model. This can probably be catered for by noting that the limit has been set by the 'best practices current at the time' and that any new methods which are believed to improve best practice should correspondingly improve the best achievable claim on software systems. Note however, that new testing techniques might improve the testing claim, but unless similar improvements were obtained in fault avoidance and fault tolerance the system claim would still be limited at 10^{-5}.

The model presented seems to give fairly repeatable results of appropriate orders of magnitude as demonstrated by the checks performed on it and so might be judged to be a reasonable starting point for assisting and ensuring consistency in the making of judgements about the adequacy or otherwise of system testing. However, the model should be evaluated against experience of actual systems so that it can be appropriately modified or discounted. While it is subjective, the method suggested is auditable and has the benefit of moving as much as possible of the quantitative element into an area which can be assessed against the performance of the model in practice, and thus reduces the level of subjective judgement required from individual analysts to a minimum. These judgements must be justified and there is scope to make an analyst choose appropriate compliance factors using guidelines such as those given for structure based testing or by the development of objective measures for each of the categories. This idea may reduce the degree of independence necessary for this assessment of the testing processes which it is suggested should otherwise conform to that given in the IEC 1508 standard for software validation reproduced in Table 9-7.

Table 9-7 : Degree of Independence of Assessment

Degree of Independence	SIL 0	SIL 1	SIL 2	SIL 3	SIL 4
Independent Company	–	R	HR	HR	HR
Independent Department	R	R	R	HR	HR
Independent Persons	R	R	R	NR	NR

R = Recommended
HR = Highly Recommended
NR = Not Recommended

This table is reproduced from [6]; © British Standards Institution.

It is possible that the use of a test regime assessment model might allow the HR for Independent Company to become R for SIL2 systems.

The scarcity of data to support software assessment means that there is uncertainty in the results of any analysis using the suggested model. However, it is believed that the conservative choice of factors has ensured that this method produces conservative results and that therefore the sensitivity of the results to slightly different choices of input data should not be a significant issue as optimistic results should not be produced.

The model currently applies to protection systems but could be extended to cover continuous control systems with failure frequencies by the use of a suitable multiplier (e.g. 10^{-4}).

The model retains the benefit inherent in the draft IEC standard that the developer is left to decide to a large extent what testing to perform and so can choose the most appropriate methods that accomplish their needs/experience.

It is considered that a necessary prerequisite for the use of this general approach is that basic quality concerns should have been addressed, such as:

- there are standards and procedures for testing with supervision to ensure that they are applied
- all tests and results are adequately documented
- there are procedures for corrective action to suitable standards if faults are detected during testing
- test documentation is reviewed prior to and after testing
- documentation and code should be subject to configuration management.

It should be noted that a similar approach is possible for the development process assessment in respect of fault avoidance and primary factors to be considered might be quality assurance, configuration management, design methodology, formal methods and simplicity. For example, the tables from the draft IEC standard which might be relevant cover lifecycle issues and documentation, software requirements specification, design and development, quality assurance and maintenance. Fault tolerance is a more difficult area and it is considered that less weight might be given to this factor than to fault avoidance and detection, although the software architecture table in the IEC standard might be a suitable starting point for an assessment.

There are four strands to safety arguments:

- credible arguments about the adequacy of the requirements specification in terms of the hazards to be addressed
- credible arguments, obtained by analytical means, about the ability of the implementation and the design to satisfy the safety requirements
- empirical evidence, obtained by dynamic testing, of the system's ability to satisfy the safety requirements
- credible arguments about the adequacy of the development process (including the humans involved) in terms of its ability to avoid the introduction of faults, and to detect and remove faults.

It is emphasised that the method presented here is largely concerned with only the third and fourth of these strands and that the arguments presented above would represent only a partial contribution to any safety case. Empirical evidence, which is provided both by the functional and environment elements of the model as well as by the random input/reliability growth testing required, is useful because it reflects the performance of the hardware and software (system and application) and so provides information about the total system. Some of the techniques covered by the model can also contribute to the first and second strands. For example, it has been suggested that plant modelling and test cases derived from hazard analyses can be used to check the safety specification and analytic methods such as response timings and memory constraints might be relevant to the second strand. It should also be emphasised that the value obtained from the model can in no way be considered the reliability actually achieved or probably achieved, but simply provides an element of confidence that testing consistent with the claimed reliability has been performed. It thus contributes to the final aim of giving confidence that the software has sufficient integrity in the context of overall system integrity/reliability.

9.5 Evidence to Support the Test Regime Assessment Model

The model has been checked to ensure that it gives reasonable results for actual systems where the safety authorities have made a judgement about the adequacy of the software with respect to the required system integrity. One check used a protection system for which the appropriate safety authority have accepted that the supplier has presented 'a suitable basis for the safety justification of the software'. This protection system is based on a GEM80 PLC and the software (application and system) is considered not to compromise the calculated hardware integrity of 1×10^{-2} pfd. The testing involved functional testing (with an emphasis on protection features), simple static analysis, and random input testing. Testing was performed by independent persons and was carried out on the hardware which was to be installed. This information when entered in the model gives a result of 1×10^{-2} pfd. A great deal of information about the previous user applications and known problems about the system software was obtained from the manufacturers and enabled the system software testing to be considered to justify a claim at the same level as that for the application software.

A second check used the primary protection system (PPS) at Sizewell B, which has a target pfd of 1×10^{-4}. The NII have accepted that in principle the technology is capable of being engineered to this level of reliability [75]. The NII 'special case procedure', which if comprehensively met will constitute a reliability demonstration at this level, has been outlined by outlined in *A Synopsis of Papers* presented at the RAE [75]. The testing elements include verification and validation (V&V) by a separate team within the design company supported by static and dynamic testing tools (MALPAS and LDRA Testbed). System testing includes installation of a prototype in a typical operating environment and exercising the system through frequent fault sequences, and also the final system will undergo a year long site test. In this case the model gives a result of 2×10^{-4} pfd although it must be noted that the testing detail entered in the model is largely guesswork based on the brief descriptions given by Hunns and Wainwright. No information on the approach adopted for system software is available and this aspect has been ignored, however it can easily be seen that if a track record approach was to be relied upon then it would have to be extensive and relevant. It is noted that a reverse compilation exercise has been performed to ensure that no errors have been introduced by the compiler.

A limited survey of projects was carried out by questionnaire, in which the Test Regime model was used to predict a reliability figure. This was compared with the probability of failure figure or Integrity Level claimed in safety arguments. A summary of the results is present in Table 9-8.

Table 9-8 : Evaluation of Questionnaire Responses Using the Model

Questionnaire Number	Application Software		System Software		Notes
	Predicted	Claimed	Predicted	Claimed	
1	5×10^{-2}	$\mathbf{1 \times 10^{-4}}$	-	-	2&3
2	7×10^{-3}	1×10^{-2}	-	-	1
3	2×10^{-3}	IL2	1×10^{-2}	IL2	1&2
4	4×10^{-4}	$\mathbf{1 \times 10^{-4}}$	-	-	1&2
5	3×10^{-3}	$\mathbf{1 \times 10^{-4}}$	-	-	2&4
6	8×10^{-5}	$\mathbf{1 \times 10^{-5}}$	-	-	1&2

(claims which are in bold have been formally validated/licensed by an external organisation)

Notes:

1 Formal methods used.
2 Probabilistic testing (mostly based on operational data).
3 Predicted level reflects only additional testing performed by the questionnaire respondent on a consultancy basis. The extent of the 'traditional' testing activities performed by the system designers/manufacturers is not known and cannot be accounted for.
4 There are problems in the assessment of 'not applicable' responses for some testing techniques in this questionnaire. It is considered likely that the prediction might be 1×10^{-3} or better if further investigation were to be taken.

The results in Table 9-8 are shown symbolically in Table 9-9 to illustrate the match between claimed and predicted probability of failure.

Table 9-9: Symbolic Representation of Questionnaire Response

Questionnaire	IL0 1×10^{-1}	IL1 1×10^{-2}	IL2 1×10^{-3}	IL3 1×10^{-4}	IL4 1×10^{-5}
1		p		C	
2		c p			
3		<c p c>			
4				p c	
5				p C	
6					p C

c – Claimed integrity by developer
C – Claimed integrity accepted by validator/licensor
p – Predicted integrity using the formula

The model has also been checked with the testing recommendations at different integrity levels from a number of safety-related standards. A summary of the results is given in the following table.

Table 9-10: Model Results Using the Recommendations from the Standards

Standard	IL1	IL2	IL3	IL4	Note
IEC [6]	7×10^{-2}	4×10^{-3}	3×10^{-4}	2×10^{-4}	1&3
RTCA [8]	7×10^{-2}	1×10^{-2}	6×10^{-3}	2×10^{-3}	2&3
BRB/LU/RIA 23 [9]	4×10^{-2} (2×10^{-2})	2×10^{-2} (4×10^{-3})	3×10^{-3} (2×10^{-4})	2×10^{-3} (2×10^{-4})	4
MoD DS55 [13]	-	-	-	6×10^{-4}	3

Notes:

1 IEC HR techniques used. No R techniques included.
2 RTCA does not cover testability (min factor 0.08).
3 These standards do not consider those testability issues (min factor 0.5).
4 RIA does not really cover environment (min factor 0.1) or testability (min factor 0.08). It has been produced as a supplement and the results have also been calculated assuming IEC recommendations in these areas are followed. These results are given in brackets.

9.6 Guidance

The *test regime assessment model* offers a systematic method for justifying a particular set of test methods. It is recommended that safety-related systems are designed, implemented and tested with reference to this *test regime*. The role of the *test cost factor* is to provide a check that the *test regime* is achievable within economic constraints. A scheme in which the two procedures may be employed is summarised below:

- **Tender**
1. The developer should determine the integrity level.
2. A suitable *test regime* should be indentified using *test regime* assessment model.
- **Top level design**
3. The estimates of *test expenditure* should be refined during top level design.
4. The *test regime* 'score' should be revised in the light of design decisions.
- **Contract completion**
5. Design expenditure, *test expenditure* and development methods should be recorded.
6. TCF following contract completion should be calculated and used as feedback for later projects.

Steps 1, 2 and 3 above should be regarded as iterative choosing, if possible, alternative architectures or features which have been shown in the past to reduce test expenditure.

References

1. Dijkstra. *Disciplines of Programming*. Prentice Hall.
2. Littlewood B. *The need for evidence from disparate sources to evaluate Software Safety Control Systems*. Symposium, Bristol 1993
3. Beizer B. *Software Testing Techniques*.
4. EPSRC Safety Critical Systems Programme Project IED4/1/9301, *Network Programming for Software Reliability* Managed by University of Exeter. Professor D. Partridge
5. *HSE Guidelines for the use of Programmable Electronic Systems in Safety-Related Applications*. 1987
6. IEC 1508, parts 1 to 7, *Functional Safety: safety related systems*. Draft, 1995
7. IEC 880, *Software for Computers in the Safety Systems of Nuclear Power Stations*. 1986
8. RTCA/DO-178B, *Software Considerations in Airborne Systems and Equipment*. December 1992
9. BRB/LU LTD/RIA Technical Specification No 23, *Safety-Related Software for Railway Signalling*. Consultative Document, 1991
10. CLC/ SC9XA/WG1 (draft), CENELEC *Railway Applications: Software for Railway Control and Protection Systems*.
11. ISA SP84, *Electric (E) Electronic (E) / Programmable Electronic Systems (PES) for use in Safety Applications*. Draft, 1993
12. DIN V VDE 801, *Grundsatze fur Rechner in Systemen mit Sicherheitsaufgaben (Principles for computers in safety-related systems*. January 1990
13. DEF STAN 00-55, parts 1 and 2, *The Procurement of Safety Critical Software in Defence Equipment*. Interim, April 1991
14. DEF STAN 00-56, parts 1 and 2, *Safety Management Requirements of Defence Systems containing Programmable Electronics*. Draft, February 1993

15   IGE/SR/15, *Safety Recommendations for Programmable Equipment in Safety-Related Applications*. Institute of Gas Engineers, Issue 6, December 1993

16   Magee S, Tripp L. *Guide to Software Engineering Standards and Specifications*. 1997

17   BS5760 part 8, *Assessment of the Reliability of Systems Containing Software*. This is currently in draft form (*Draft for Development 198*)

18   CONTESSE Task 12 Report, *Hazard and Operability Study as an Approach to Software Safety Assessment*.

19   Interim Defence Standard 00-58, *HAZOP Studies Guide for Programmable Electronic Systems*. Draft in preparation.

20   *Standard for Software Engineering of Safety Critical Software*. Ontario Hydro, 1990

21   Leveson N G and Harvey P R. *Analyzing Software Safety*. IEEE Transactions on Software Engineering 1983; Vol. SE-9, No. 5

22   Myers G J. *The Art of Software Testing*. 1979

23   Myers G J. *Software Reliability*. John Wiley & Sons, 1976

24   Boehm B W. *Software Engineering Economics*. Prentice Hall, 1981

25   Cullyer W J, Goodenough S J, Wichmann B A. *The choices of computer languages for use in Safety Critical Systems*. Software Engineering Journal 1991; 6:51-58

26   *The Statemate Approach to Complex Systems*. i-Logix, 1989

27   *The Semantics of Statecharts*. i-Logix, 1991

28   Halstead M H. *Elements of Software Science*. Elsevier North Holland, New York, 1977

29   McCabe T J. *A Complexity Measure*. IEEE Transactions on Software Engineering 1976; Vol. SE-2; No. 4

30   Pyle I C. *Developing Safety System – A Guide Using Ada*. Prentice Hall International.

31   CONTESSE Task 8 Report, *An Integrated Platform for Environment Simulation*.

32   Bache R and Mullerburg M. *Measures of testability as a basis for quality assurance*. Software Engineering Journal, March 1990; 86-92

33   CONTESSE Task 4 Report, *Software Testing via Environmental Simulation*.

34   Balci O. *Credibility Assessment of Simulation Results: The State of the Art*. In: Proc. Conference on Methodology and Validation, 1987, Orlando, Fla., pp 19-25.

35 Carson J S. *Convincing Users of Model's Validity Is Challenging Aspect of Modeler's Job*. Ind. Eng. June 1986; 18:74-85.

36 Sargent R G. *A Tutorial on Validation and Verification of Simulation Models*. In: Proc. 1988 Winter Simulation Conference, San Diego, Calif., pp 33-39.

37 Naylor T H and Finger J M. *Verification of Computer Simulation Models*. Management Sci. 1967; 5:644-653.

38 Goodenoug J B and Gerhart S L. *Toward a theory of test data selection*. IEEE Transaction on Software Engineering, June 1975; Vol.SE_3

39 Weyuker E J. *Assessing test data adequacy through program inference*. ACM Transactions on Programming Languages and Systems, October 1983; Vol. 5, 4:641- 655.

40 Ould M A and Unwin C (eds). *Testing in Software Development*. Cambridge University Press, 1986.

41 Basili V and Selby R W. *Comparing the effectiveness of software testing strategies*. IEEE Transactions on Software Engineering, December 1987; Vol. SE-13,12:1278-1296

42 Myers G J. *A controlled experiment in program testing and code walkthrough/inspections*. Communications of the ACM, September 1978; Vol. 21, 9:760-768.

43 CONTESSE Task 9 Report, *Software Testing Adequacy and Coverage*.

44 Duran S C and Ntafos J W. *An evaluation of random testing*. IEEE Transaction on Software Engineering, July 1984; Vol. SE_10, 4:438-4444.

45 Gourlay J. *A mathematical framework for the investigation of testing*. IEEE Transaction on Software Engineering, November 1983; Vol.SE_9, 6: 686-709

46 Weyuker E J. *Axiomatizing software test data adequacy*. IEEE Transaction on Software Engineering, December 1986; Vol. SE_12, 12:1128-1138

47 Parrish A and Zweben S H. *Analysis and refinement of software test data adequacy properties*. IEEE Transaction on Software Engineering, June 1991; Vol. SE_17, 6:565-581.

48 Parrish A and Zweben S H. *Clarifying Some fundamental Concepts in Software Testing*. IEEE Transactions on Software Engineering, July 1993; Vol. 19, 7:742-746.

49 Zhu H and Hall P A V. *Test data adequacy measurement*. Software Engineering Journal, Jan. 1993; Vol. 8, 1:21-30.

50 Zhu H, Hall P A V, May J. *Understanding software test adequacy -- An axiomatic and measurement approach*. In: Proceedings of IMA Conference on Mathematics of Dependable Systems, Royal Holloway, Univ. of London, Egham, Surrey 1st-3rd Sept. 1993

51 Clarke L A, Podgurski A, Richardson D J, Zeil S J. *A formal Evaluation of data flow path selection criteria*. IEEE Transactions on Software Engineering, November 1989; Vol. 15, 11:1318-1332

52 Rapps S and Weyuker E J. *Selecting software test data using data flow information*. IEEE Transaction on Software Engineering, April 1985; Vol. SE_11,4:367-375

53 Ntafos S C. *A comparison of some structural testing strategies*. IEEE Transaction on Software Engineering, June 1988; Vol. SE-14, 868-874

54 Frankl P G and Weyuker J E. *A formal analysis of the fault-detecting ability of testing methods*. IEEE Transactions on Software Engineering, March 1993; Vol. 19, 3:202- 213

55 Miller W M, Morell L J, Noonan R E et al. *Estimating the probability of failure when testing reveals no failures*. IEEE Trans. on Software Engineering 1992; Vol. 18 ,1

56 Littlewood B and Strigini L. *Validation of ultra-high dependability for software-based systems*. Comms. of the ACM 1993

57 Parnas D I et al. *Evaluation of safety critical software*. Comms of the ACM 1990; Vol 33:6

58 Howden W D. *Functional Program Testing and Analysis*. McGraw-Hill, 1987

59 Poore J H, Mills H D, Mutcher D. *Planning and certifying software system reliability*. IEEE Software, January 1993

60 Ehrenberger W. *Probabilistic techniques for software verification in safety applications of computerised process control in nuclear power plants*. IAEA-TECDOC-58, February 1991

61 Littlewood B. *Forecasting Software Reliability*. In: Bittanti S (ed). *Software Reliability Modelling and Identification*. Springer-Verlag, 1988

62 Mills H D, Dyer M, Linger R C. *Cleanroom Software Engineering*. IEEE Software, September 1987

63 Brocklehurst S, Chan P Y, Littlewood B, Snell J. *Recalibrating software reliability models*. IEEE Trans. on Software Engineering, 1990; Vol.16,4

64 Littlewood B. *Software reliability models for modular program structure*. IEEE Trans. on Reliability, October 198; Vol.30:313-320

65	Mellor P. *Modular structured software reliability modelling in practice*. 4th European Workshop on Dependable Computing, Prague 8th-10th April 1992
66	Laprie J C, Kanoun K. *X-ware reliability and availability modelling*. IEEE Trans. on Software Engineering 1992; Vol. 18,2
67	Knight J C, Leveson N G. *An experimental evaluation of the assumptions of independence in multiversion programming*. IEEE Trans. on Software Engineering 1986; Vol. 12,1
68	Butler R W, Finelli G B. *The infeasibility of experimental quantification of life critical software reliability.* In: Procs. ACM SIGSOFT `91, Conference on Software for Critical Systems, New Orleans, Dec. 4th-6th 1991
69	May J H R, Lunn D. *A model of code sharing for estimating software failure on demand probabilities*. Tech. rep., Depts. of Computing and Statistics, May 1993
70	Casella G, Berger R L. *Statistical Inference*. Wadsworth and Brooks/Cole, Belmont Calif, 1990
71	Hamlet R G. *Are we testing true reliability?* IEEE Software, July 1992
72	May J H R and Lunn D. *New Statistics for Demand-Based Software Testing*. Information Processing Letters 1995; Vol53
73	CONTESSE Task 11 Report, *Practical Specification of a Simulation of a Probabalistic Environment for a Reactor Protection System.*
74	Hughes G, Pavey D, May J H R et al. *Nuclear Electric's Contribution to the CONTESSE Testing Framework and its Early Application.* Procs. of the 3rd. Safety-Critical Systems Symposium, Brighton February 1995
75	Hunns D M and Wainwright N. *Software-based Protection for Sizewell B: the Regulator's Perspective*. Nuclear Engineering International, September 1991
76	Musa J D. *Operational Profiles in Software Reliability Engineering*. IEEE Software, March 1993.
77	Brooks F P. *The Mythical Man Month*. Addison Wesley, 1975
78	IEEE. *Guidelines for Assuring Testability.* 1988, ISBN 0-86341-129-0
79	BSI DD198 :1991. *Guide to Assessment of reliability of systems containing software.*
80	DOD-STD-2167A. *Military Standard, Defence System Software Development.*

Appendix A
Summary of Advice from the Standards

This appendix contains a review of a range of standards and guidelines for the development of safety-related systems, and the advice given on testing. A discussion on the current industry practices and the use of the standards and guidelines can be found in Chapter 1.

HSE Guidelines [5]	
Status and Content Overview	*Summary of Testing Advice*
Guidance document for the development and use of programmable electronic systems in safety-related applications. Provides checklists to aid software safety assessment	Very limited guidance on testing given although some test coverage criteria suggested. Mathematical models of software reliability are considered to have limited usefulness in the context of PES. Static analysis is considered desirable.

IGE/SR/15 [15]	
Status and Content Overview	*Summary of Testing Advice*
Draft application specific guidelines for the gas industry (ranging from offshore installations to domestic appliances) on the use of programmable equipment in safety-related applications. Based on HSE Guidelines.	Very limited guidance on testing – little more than that already provided in the HSE Guidelines. Advises that the use of formal methods be considered, particularly for complex systems. Dynamic testing regarded as essential at all levels of development, but no direct advice on methods provided. Use of static analysis encouraged with specific types of analysis available and the languages they are applicable to discussed.

Drive Safely	
Status and Content Overview	*Summary of Testing Advice*
Proposal for European Standard for the development of safe road transport informatic systems.	Provides limited guidance on test objectives. Guidance varies according to the required safety integrity level.
	Suggests static analysis and testing should be carried out on *high* integrity software. For *very high* integrity software, formal proofs are required and test coverage must be justified. Black and white box testing required at all levels.

IEC 1508 [6] – Part 3 Software Requirements

Status and Content Overview

Draft international standard on suitable techniques/measures to be used during development of safety-related software

Provides recommendations varying by required safety integrity level

Developer can be selective about techniques/measures chosen so long as the choice is justified

Summary of Testing Advice

Recommends test methods and techniques to be adopted for different levels of safety integrity.

The test methods are:

- formal proof – recommended for levels 2 and 3, highly recommended for level 4

- static and dynamic analysis – recommended for level 1, highly recommended for level 2 and above

- probabilistic testing – recommended for levels 2 and 3, highly recommended for level 4

- functional and black box testing – highly recommended for all levels

- performance testing – recommended for levels 1 and 2, highly recommended for levels 3 and 4

- interface testing – recommended for levels 1 and 2, highly recommended for levels 3 and 4

- simulation/modelling – recommended for levels 1 and 2, highly recommended for levels 3 and 4

For some test methods a set of suitable test techniques, graded for appropriateness by integrity level, are also provided.

IEC 1508 [6] – Part 2 Requirements for electrical /electronic / programmable electronic systems

Status and Content Overview	Summary of Testing Advice
Draft international standard setting out a generic approach to the development of electrical/ electronic/ programmable electronic systems (E/E/PES) that are used to perform safety functions. Provides recommendations on techniques/measures appropriate to the development of E/E/PES of differing levels of System Integrity. Developer can be selective about techniques/measures chosen so long as the choice is justified.	Recommended test techniques/ measures include: • simulation – recommended for levels 3 and 4 • functional testing – highly recommended for all levels • expanded functional testing – recommended for level 1, highly recommended for levels 2 to 4 • functional testing under environmental conditions – highly recommended for all levels • statistical testing – recommended for levels 3 and 4 • static and dynamic analysis – recommended for levels 2 to 4 • surge immunity testing – highly recommended for level 4 • systematic testing – recommended for level 1, highly recommended for levels 2 to 4 • black box testing – recommended for all levels • interference immunity testing – highly recommended for all levels

RIA 23[9]

Status and Content Overview

Draft proposal for railway industry standard for the procurement and supply of programmable electronic systems for use in railway applications.

Intended to be used in conjunction with IEC 1508

Summary of Testing Advice

Recommended test techniques/measures include:

- reliability growth modelling – recommended for levels 1 and 2, highly recommended for levels 3 and 4

- error guessing – highly recommended for levels 1 to 4

- error seeding – recommended for levels 3 and 4

- boundary value analysis – highly recommended for levels 1 to 3, mandatory level 4

- test coverage (including multiple condition coverage) – recommended for levels 1 and 2, highly recommended for levels 3 and 4

- functional testing – mandatory for all levels

- static code analysis – recommended for levels 1 and 2, highly recommended for level 3, mandatory for level 4

- transaction analysis – recommended for levels 1 to 3, highly recommended for level 4

- simulation – recommended for all levels

- load testing – highly recommended for all levels

- performance monitoring – recommended for levels 1 to 3, highly recommended for level 4

CENELEC [10]	
Status and Content Overview	*Summary of Testing Advice*
Draft European standard specifying procedures and technical requirements for the development of programmable electronic systems for use in railway control and protection applications.	Provides recommendations varying by required safety integrity levels. Required techniques for level 1 are the same as for level 2, techniques for level 3 are the same as for level 4. Based on IEC 1508 and RIA 23 • functional and black box testing – highly recommended for all levels • performance testing – highly recommended for levels 3 and 4 • interface testing – highly recommended for all levels • formal proof – recommended for levels 1 and 2, highly recommended for levels 3 and 4 • probabilistic testing – recommended for levels 1 and 2, highly recommended for levels 3 and 4 • static and dynamic analysis – highly recommended for all levels

DEF STAN 00-55 [13]

Status and Content Overview	Summary of Testing Advice
Interim MoD standard for the procurement of safety-related software in defence equipment. Mandates specific software development methods to be followed.	Mandates the use of formal methods, dynamic testing and static path analysis as verification and validation techniques. Formal methods are to include formal specification and design, and verification via formal proof or rigorous argument. Mandates the use of specific types of static analysis (e.g. control flow analysis) and specific dynamic test coverage criteria (e.g. all statements). Executable prototype to be produced derived from the formal specification and used to compare results of testing integrated software via a test harness. A simulator may be used if it is not practical to validate a system in its final system environment, but it is noted that 'a simulator is a complex system and its environment may be more complicated that the software under test'. Resource modelling and performance analysis are to be carried out. Validation is to include sufficient testing 'to obtain a statistically significantly measure of the achieved reliability of the system'.

Copies of Defence Standards DEF STAN 00-55 and 00-56 can be obtained free of charge from:

Directorate of Standardization
Ministry of Defence
Room 1138
Kentigern House
65 Brown Street
Glasgow
G2 8EX

Further information can be obtained from the Def Stan web site:

    http://www.dstan.mod.uk

DEF STAN 00-56 [14]

Status and Content Overview

Draft MoD standard defining the safety management requirements for defence systems containing programmable electronics.

Mandates that design rules and techniques for safety systems development are to be defined and approved prior to implementation.

Supporting guidance document provides some illustrative examples of topics to be considered when determining appropriate rules/techniques for computer based applications. Guidance document refers to IEC 1508 for further guidance on techniques.

Summary of Testing Advice

Illustrative examples provided in guidance document show:

- dynamic testing required for all levels of safety integrity
- static testing required for safety integrity levels 3 and 4, optional for levels 1 and 2.

DOD-STD-2167A [80]

Status and Content Overview

US military standard for defence system software development.

Leaves contractor responsible for selecting software development methods that best support the achievement of contract requirements and also for performing whatever safety analysis is necessary to ensure that the potential for hazardous conditions is minimised.

Summary of Testing Advice

No guidance provided on specific test measures/techniques.

Limited guidance on test coverage criteria

Guidance concentrates on the documentation to be produced.

AG3

Status and Content Overview	Summary of Testing Advice
Assessment guide for the software aspects of digital computer based nuclear protection systems	Guidance provided on the level of test coverage required.
	Static and dynamic testing to be undertaken by both supplier and independent assessor.
	Where practical, requirements animation to be carried out.
	Credit may be claimed for use of formal specifications.
	Random and back-to-back testing suggested as powerful test techniques.
	Extended period of on-site testing of full system required. During this testing a probabilistic estimate of the software reliability is to be computed (no guidance provided on how) for comparison against judged reliability value based on development methods.

IEC 880 [7]

Status and Content Overview	Summary of Testing Advice
International standard defining the requirements for the generation and operation of software used in the safety systems of nuclear power stations.	Both statistical and systematic testing are to be carried out. Detailed coverage criteria are specified for each.
	Automatic test tools are to be used as much as possible to derive test cases.
Minimum standard recommended by the UK Nuclear Installations Inspectorate.	Static and dynamic simulation of inputs are to be used to exercise the system in normal and abnormal operation.

Ontario Hydro Standard [20]	
Status and Content Overview	*Summary of Testing Advice*
Canadian standard specifying requirements for the engineering of safety-related software used in real-time control and monitoring systems in nuclear generating stations. All requirements of the standard must be met.	Statistical and systematic testing are to be carried out to ensure adequate and predefined test coverage is achieved. A quantitative reliability value is to be demonstrated using statistically valid, trajectory based, random testing. Test case requirements for this testing are specified. Detailed test objectives and coverage criteria are defined for unit, subsystem, validation and reliability testing. Dynamic testing and simulation are recognised verification techniques.

RTCA-178B [8]

Status and Content Overview

Document providing the aviation community with guidance for determining that the software aspects of airborne systems and equipment comply with air worthiness requirements.

Guidance is not mandated but represents a consensus view.

Summary of Testing Advice

Suggests use of target computer emulator or host computer simulator during testing. Emulator/simulator must be assessed as having achieved a predefined level of safety integrity. The differences between simulator/emulator and the target system and their effects on the verification process to be considered.

Emphasises the use of requirements based test methods, in particular hardware and software integration testing, software integration testing and low-level testing. The objectives of each of these test methods is defined and typical errors which should be revealed by the method identified. Guidance is provided on test case selection covering normal and abnormal range test cases.

Requirements based test coverage analysis and software structure coverage analysis is to be carried out. Guidance on coverage criteria is provided with the level of coverage required varying by integrity level.

Formal methods, exhaustive input testing and multi-version software considered potential 'alternative methods' for software verification. Some guidance is provided on their use, but not in the main guidance sections of the document as they are felt to be currently of 'inadequate maturity' or 'limited applicability for airborne software'.

Guidance on software reliability models explicitly excluded as it is believed that currently available methods 'do not provide results in which confidence can be placed to the level required'.

Bibliography

The documents listed below were produced by the CONTESSE project and are available (for a nominal cost, to cover printing, postage and packaging) from Mike Falla at:

MFR Software Services Limited
35 Benbrook Way
Macclesfield
Cheshire
SK11 9RT

Information on the CONTESSE project can also be found on the DTI website:

`http://www.augusta.co.uk/safety`

The partner in the consortium responsible for leading the work described is also given.

Id	Title
B1	*Survey of Current Simulation and Emulation Practices*
	This report presents a brief survey of the current simulation and emulation techniques used in the development of high integrity systems. The responses to these questionnaires are analysed and presented to provide an overview of current industry practices. Two case studies (from the avionic and nuclear industries) provide a more detailed study of the techniques currently employed in those industries.
	Lead Partner: University of Warwick
B2	*Definition of Environment Lifecycle Models*
	This report discusses the lifecycle for the development of real-time high integrity systems. A dual lifecycle is proposed, which places increased emphasis on environment simulation as a method of dynamic testing. Methods of implementing the various components of the lifecycle are proposed that take into account the integrity level and complexity of the target system.
	Lead Partner: University of Warwick
B3	*Analysis of Software Testing Quality*
	This report covers:
	• a review of the test quality analysis and targets, both as used by CONTESSE partners in general practice and as documented in current research
	• a review of the use of tools in support of test quality analysis
	• a categorisation of the approaches by application sector.
	Lead Partner: Nuclear Electric
B4	*Software Testing by Environmental Simulation*
	This report is concerned with the theoretical foundations of software testing through environment simulation. The report covers:
	• formal descriptions of software and associated environment simulation, within which the problems of test adequacy, test case generation and software integrity estimation can be formulated
	• the theoretical basis of test adequacy, and of software integrity, measurement when environment simulation is used to test software
	• the relationship between test adequacy and software integrity when environment simulation is used to test software.
	Lead Partner: Nuclear Electric

B5	*Testability of the Design Features of Programmable Electronic Systems*
	This report uses a combination of experience gained with development of existing systems, and primitive control and data flow models of alternative design solutions to attempt to formulate a scale of testability of design features. The features of a system's interfaces with its environment, which make the construction of effective simulators difficult, is also analysed in a quantitative manner.
	Lead Partner: Lloyds Register
B6	*Requirements for Test Support Tools*
	This report defines environment lifecycle models and the requirements for the effective use of simulators. The report also analyses the overall development process for simulators and identifies those areas which would benefit from tools which directly contribute to this aspect of the development. Requirements for tools are defined to a level sufficient to evaluate those currently available.
	Partner: NEI
B7	*Computer Aided Software Testing: Tool Survey*
	This report provides the results of a survey of tools that may assist in testing software. Most of these tools are sold as part of an integrated development environment and some are also sold as stand-alone tools. The information contained within this report has been extracted from the tool vendors literature.
	Lead Partner: University of Warwick
B8	*Computer Aided Software Testing: Tool Survey and Evaluation*
	This report presents and evaluates the result of a survey of software and PLC test tools. Four software test and simulation tools were selected, "McCabe", "STATEMATE", "AdaTEST", and "MatrixX and SystemBuild".
	Lead Partner: University of Warwick
B9	*Tool Support for the CONTESSE Framework*
	This report discusses aspects of tool integration and identifies a set of tools to support the CONTESSE Framework. This set is compared with the outcome of work carried out earlier in the project when specific test support tools were identified. The likelihood of obtaining tools to support the Framework is considered.
	Lead Partner: NEI

B10	*CONTESSE Test Handbook*
	This report contains a compilation of the results of all the tasks undertaken in the CONTESSE project. The report is in four sections:
	• Background
	• Test Strategy and Methods
	• Tools and Simulators
	• Case Studies
	This report contains the detailed information on which this book is based.
	Lead Partner: BAeSEMA
B11	*Fire and Gas Protection System Case Study*
	This report describes the particular features of offshore fire and gas protection systems that may benefit from of the guidelines in the CONTESSE Test Handbook, and describes the results of applying the guidelines, particularly those relating to:
	• safety cases
	• aspects of environmental lifecycle models relating to construction of the environment simulator
	• testability.
	Lead Partner: G P-Elliot
B12	*Application of the CONTESSE Framework to a Reactor Protection System*
	This report investigates the feasibility of including reactor protection software in a quantified overall nuclear power plant safety argument. To support the investigation a mathematical model was developed and a statistical software testing experiment conducted, using a simulated environment, to estimate the probability of failure on demand of protection software.
	Lead Partner: Nuclear Electric
B13	*Review Implementation against Framework and Test Framework on Model*
	This report reviews the applicability of the CONTESSE Test Handbook to the certification of safety-related programmable electronic systems within the civil aviation industry. It is structured to facilitate comparison of the CONTESSE guidelines and RTCA DO-178B, the aviation guidelines for software in airborne systems and equipment.
	Lead Partner: Lucas Electronics

B14	*PTSE Case Study Final Report*
	This report discusses whether a system's claimed integrity can be improved by the execution of further tests. It reports on a study carried out using a duplicate of an existing reactor protection system subjected to a series of pseudo random tests. It concludes that the results from such tests can support an argument for an improved system integrity.
	Lead Partner: NEI
B15	*The Application of Simulation to Timing Constraints*
	This report discusses simulation methods, techniques and tools for improving the specification, verification and testing of real-time software timing constraints at an early stage during the software development lifecycle. A comparison of commercially available tools against a functional definition of an 'ideal' tool is provided.
	Lead Partner: Rolls-Royce
B16	*The application of the CONTESSE Framework to a Marine Roll and Stabilisation System.*
	This report describes a proposal for a Rudder Roll Stabilisation System, and with reference to the CONTESSE Framework considers the validation and testability of the proposed system.
	Lead Partner: Lloyds Register

Index

accessibility 66
accumulation metrics 135
accuracy
 of environment simulation 134, 136
 evaluation 136
 of real number system 133
 of reliability estimation 146
 of simulation 137–138
 of simulators 132, 141
Ada 79, 81, 94, 98, 105, 109, 118
AdaTEST 105–106, 116
adequacy criteria
 boundary analysis 149
 control flow based 154
 fault based 149
 functional analysis 149
adequacy test 143
ALARP 42
algorithm
 heuristic 91
 rate monotonic 86, 89
 scheduling pre-emptive 86
analysis
 boundary 147
 boundary value 180, 182–183, 205
 causal net 56
 code coverage 104
 complexity 22, 104
 control flow 95, 183
 data flow 183
 design 41
 dynamic 203–204, 206
 failure modes and effects 38
 hazard 35, 43–44, 54, 102, 127, 184–185
 information flow 183
 mutation 150
 performance 207
 risk 33, 41, 44, 102
 safety 208
 sensitivity 43
 sneak circuit 183
 static 31, 39, 79, 100, 121, 149, 176, 186, 201, 203–204, 206–207
 static path 207
 structured 103
 timing 100
animation 182
 of requirements 19
 of specifications 61
animators 176
architecture target 97
As Low As Reasonably Practicable (ALARP) 42
assessment
 independent 25
 process 2
 product 2
 quantitative 25
 safety 25
 test cost 171
availability 51
avalanche testing 182

back-to-back testing 209
behavioural constraints 83
behavioural model 125
bespoke architecture approach 14
black box testing 31, 40, 111–112, 147
boundary analysis
 adequacy criteria 149

boundary value analysis 180, 183, 205
branch coverage 146
brute-force searching 91

C 98, 105
C++ 81, 94
causal net analysis 56
Cause Consequence Analysis (CCA) 57
cause consequence diagrams 182
cause-effect
  descriptions 40
  modelling 35
  relationships 35
CCA 57
channel diverse 7
checklists 183
cleanroom technique 157
code
  coverage 22
  coverage analysis 104
  inspection 21, 37
combinator 17
competence of personnel 41
compilers 82, 109
  Ada 98
complexity 22, 61, 76, 78, 103, 140, 166, 180
  analysis 104
  McCabe's metric 79
component environment simulators 127
computation errors 149
confidence interval 164
configuration management 111, 121
control flow 96, 98
  analysis 21, 95, 183, 207
  based adequacy criteria 154
  data 21
  test data adequacy criteria 148
control loop performance 60
control path 158
control system 4, 59, 160
controllability 66, 163
cost
  drivers 175
  of testing 3, 146
coverage
  branch 146
  of code 144
  criteria 154, 209
  data path 158
  decision 117
  exception 118
  path 145
  statement 118, 145
  structure 148
coverage based testing 21
criterion structure coverage 146
cyclic scheduler 90

data flow 64, 110
  analysis 183
  criteria 148, 154
  structure 154
  test data adequacy criteria 148
data path coverage 158
decision coverage 117
decompositional modelling 157
demand based models 156
demand based SST 160
design
  analysis 41
  complexity 175
  data flow based 20
  data structure based 20
  object oriented 20
  reviews 183
  standards 42
  test case 40
designated architecture approach 14
deterministic static cyclic scheduling 86
diagnostic coverage 10, 15
diverse
  channels 7
  hardware 72
  systems 7
diversity 7
document reviews 37
domain error 149
drivers cost 175
dual lifecycle model 49
dynamic
  analysis 204, 206
  memory allocation 94
  simulation 209
  test 120
  test coverage criteria 207
  testing 21, 25, 64, 100, 110, 201, 208–209
  testing tools 191

Index   221

embedded software 9
emulation 184
emulator 126, 211
end-to-end deadline times 93
environment 188
　lifecycle 54
　simulation 105, 125, 132, 139,
　　155, 159
　simulator 31, 50, 52, 141, 170
environmental and operating
　　conditions 38
environmental model 125
environmental modelling 104
equivalence
　classes 182
　in non-determinism 134
　in timing and time delay 134
　of two environments 130
equivalence relation approach 125,
　　132–133
error 6
　domain 149
　guessing 182–183, 205
　human 1, 57
　random 178
　seeding 150, 182, 205
　systematic 178
　timing 85
error based criteria 149
error computation 149
estimation of reliability 146
ETA 56
evaluation accuracy 136
event
　space 130, 136
　trapper 64
Event Tree Analysis (ETA) 56
evidence
　process 36
　product based 36
exception coverage 118

Fagan inspections 183
failure 6
　random 24
Failure Modes and Effects Analysis
　　(FMEA) 38
Failure Modes, Effects and Criticality
　　Analysis (FMECA) 55
failures 1
fault 6

fault based
　adequacy criteria 149–150
fault propensity 158
Fault Tree Analysis (FTA) 38, 56
finite state machines 61
flow graph based 148
flow graph model 148
FMEA 38
FMECA 55
formal
　methods 31, 109, 163, 192, 201,
　　207, 211
　proof 31, 202–203, 206
　semantics 110
　specification 175, 207, 209
frequency response requirements 60
FTA 38, 56
functional
　analysis 39
　independence 37
　testability 66
　testing 21, 31, 64, 179, 181, 183–
　　184, 188, 204–206
functional analysis
　adequacy criteria 149

global data areas 79

Halstead's software science 118
Hansen's software complexity 118
hard real-time systems 85, 89, 100,
　　120
hardware
　diverse 72
　interface 62
　monitors 63
hazard 1, 4, 34, 41
　identification 42–43
　log 47, 54
hazard analysis 35, 37, 42–43, 54,
　　103, 107, 127, 161, 165, 184–
　　185, 190
Hazard and Operability Study
　　(HAZOP) 55
HAZOP 38, 43, 55, 102
HEA 57
heuristic algorithms 91
human
　error 1, 57
　factors 146
　failure modes 57

Human Error Analysis (HEA) 57

independent assessment 25, 44
inference adequacy criteria 145
information flow analysis 183
information hiding 79
input
    partition testing 182
    trajectories 160
instruction prediction 98
integrated project support
    environment (IPSE) 121
integrity level 6, 33, 44, 46, 51, 107,
    109, 172, 174, 193
interface
    considerations 62
    hardware 62
    software 62
    testing 183
interrupt driven 38
interrupts 86–87, 185
IPSE 121
iteration rates 92

justification of safety claims 45

ladder logic 60
language construct 21
language subset 94
LDRA Testbed 106
lifecycle
    dual 3, 49
    environment 54
    model 26
    project safety case 48
    safety 3, 45, 54, 126
limits of safe operation 41
load testing 205
logic
    control 38
    trees 56

MALPAS 191
Markov chains 132
mathematical proof 86
Matrix-X 105–106
maximum difference metric 135
Mc Cabe's 79
McCabe Toolset 105–106
McCabe's measure 118

memory allocation dynamic 94
memory constraints 182
methods formal 31
metric
    accumulation 135
    maximum difference 135
    space 134
metric space approach 125, 132, 134
minimum time gap in input-output
    processes 93
minor cycle overload 93
model
    demand based 156
    flow graph 148
    lifecycle 26
    set theoretical 129
    stochastic process 131
    time based 156
    urn 156
modelling
    cause-effect 35
monitoring and auditing
    procedures 43
mutation adequacy 145–146, 150
mutation analysis 150
Myre's extension 118

natural language 85
network programming 7

object oriented 20
    design 77
    programming 77
object-code reading 98
observability 66
operating system multi-tasking 20
operational envelope 64
organisational
    guidelines 41
    procedures 41

partial beta factor model 178
partition testing 152, 168
Pascal 94, 105, 109
path coverage 145, 159
penalty and benefit 93
performance
    analysis 207
    constraints 83
    control loop 60

degradation 80
model 20
monitoring 205
requirements 182
step response 60
testing 31, 203, 206
periodic processes 87
Petri Nets 85
pfd 161, 179
post-conditions 22, 64
pre-conditions 22, 64
predictability 66
probabilistic testing 192, 203, 206
probability
of failure 158
of failure on demand (pfd) 161, 179
procedural parameter 79–80
process 45
assessment 2
periodic 87
simulation 182
sporadic 87
processor utilisation 89
product assessment 2
program based adequacy criterion 150
program text based 148
project safety case lifecycle 48
proof check 10
proof formal 31
protection system 4, 25, 62, 72, 155–156, 159–160, 179
prototyping 182

QA 48
qualifications and experience requirements 43
quality
assurance 25
control 25
quality assurance (QA) 48
quality requirements 101
quantitative hardware analysis 11

random 6
errors 178
failure 24
input testing 21, 177
testing 112, 148, 152, 155
rate monotonic algorithm 86, 89

real time
hard 133
soft 133
recursion prohibition 94
redundancy 6
active 6
dual 6
passive 6
triple 6
reliability 51, 122, 129, 143, 155, 157, 177, 190, 207, 209–211
growth modelling 156, 177, 181, 205
probabilistic estimate of 25
quantitative 31
repeatability 66
required integrity level 109, 154
requirements
animation 209
specification 45, 50
Requirements and Traceability Management (RTM) 113
resource
modelling 31
usage 39
utilisation 65
response timings 182
review procedures 41
risk
analysis 33, 41, 44, 163–164
assessment 33–34, 102
reduction 44
tolerable 42
tolerable level 44
risk reduction
facilities 44
measures 54
requirements (RRR) 102
techniques 41
RRR 102
RTM 105–106, 113

safety
analysis 208
argument 158
assessment 24–25, 28, 43, 48, 177, 201
assessment methodology 43
assurance 1
case 1–2, 17, 26, 33, 37, 41–42, 54, 102, 107, 139, 141, 171
compliance assessment 44

function 5, 20, 34, 54
hazard 29
integrity 1, 8, 30–31, 34, 107
integrity level 54, 123
integrity targets 44
justification 3, 20, 33, 103, 121, 125
lifecycle 3, 17, 33, 45, 49, 54, 102, 126
lifecycle phases 45
management plan 50
management system 43
objectives 25, 30
principles 42
records 44
requirements 5, 29, 33, 161
requirements specification 50
review 41
risk assessment 107
specification 190
targets 41
validation 52
validation report 53
verification 44
safety function 10
sample path 131
sampler 64
schedule pre-run-time 100
scheduler
   cyclic 90
   non-pre-emptive 86
   pre-emptive 86
   static cyclic 90
scheduling
   deterministic static cyclic 86
   pre-run-time 100
   pre-run-time (or static) 87
   run-time 87
self tests 62
sensitivity analysis 43
set theoretical model 3, 125, 129, 141
simulated annealing 90–92
simulation 22, 27, 68, 121
   accuracy 125, 132, 134, 137–138, 141
   dynamic 209
   environment 132
   static 209
   stochastic 89
simulator 3, 101, 125, 139
single-version testing 156
sneak circuit analysis 183

software
   complexity 154
   decomposition 157
   hazard analysis 37–39
   integrity 155, 157
   interface 62
   reliability 25, 168, 178
   safety 29
   safety assessments 54
   variants 81
software structure coverage analysis 211
specification of safety requirements 41
specification animators 127
sporadic processes 87
SST 155
   demand based 160
state transitions 61
Statemate 104, 106, 110
statement coverage 118, 145, 159
static
   analysis 21, 31, 37, 39, 78–79, 100, 117, 121, 149, 179, 182–183, 186, 188, 201–202, 204–206
   cyclic scheduler 90
   simulation 209
   testing 25, 120, 208–209
   testing tools 191
statistical
   inference 164
   software testing 3
   testing 204, 209
statistical inference 170
Statistical Software Testing (SST) 155
step response performance 60
stochastic
   evolution 92
   process model 3, 125, 131
   simulation 89
stop rule 146
stress testing 182
structural testing 179, 183, 188
structure coverage 148
structure diagrams 76
structure-based testing 182
structured analysis 103
structured design 76, 81
SUM 164
surge immunity testing 204
symbolic execution 183
syntactic structure 147

system
    certification 43
    diverse 7
    modularity 79
    performance 20
    reliability 172
    resources 39
    testability 171
    validation specification 50
system function calls 39
systematic errors 178
systems
    hard real-time 85
    protection 25

target system architecture 97
target system drivers 127
TCF 171
Teamwork 113
TeamworkRqt 106
TeamworkSIM 106
test
    acceleration 169
    adequacy 3, 143, 157
    adequacy measurement 155
    case selection criterion 144
    cases 22, 64, 128, 144
    cost factor 172
    coverage 26–27, 31
    coverage based 21
    coverage criteria 26
    criterion 143
    data adequacy criteria 144
    dynamic 21, 25, 27, 111, 120
    environment 52, 101
    expenditure 19, 172, 194
    functional 21
    harness 23
    methods 26, 40
    objectives 52
    plan 19
    planning 19, 108, 154
    quality 154
    random input 21, 177
    regime 22, 102–103, 172
    regime assessment 3, 171
    report 52
    schedule 112
    script 116
    specification 20, 52, 71
    static 25, 27, 120
    strategy 102
    validity 60, 79, 166
    white box 104
test adequacy criteria
    functional testing 146
    statement coverage 144
test adequacy measurement
    boundary analysis 147
test case
    design 40
    execution 182
    generation 40
    management 23, 104
    selection 145, 211
test cost assessment 171
test cost factor (TCF) 171
test coverage 202, 205, 210
    requirements based analysis 211
test coverage criteria 208
test data adequacy criteria
    control flow 148
    data flow 148
testability 3, 59, 64, 66, 75, 82, 151, 173, 179, 184–186, 188
testing
    back-to-back 209
    black box 40, 111, 203
    dynamic 25, 31, 64, 110, 208–210
    functional 64, 203–204, 206
    performance 31, 203, 206
    probabilistic 192
    random 112, 209–210
    static 31, 208–209
    statistical 204, 209–210
    systematic 209–210
    trajectory based 210
    white box 40, 111
testing approach 148
testing techniques
    boundary value analysis 182
    dynamic 177
    functional testing 182
    partition testing 168–169, 182
    random testing 155, 181
    static 177
    structural testing 182
time based models 156
time deterministic mode 20
time index set 131
time resolution 131
Timed CSP 85–86, 100
timed petri nets 86
timing 3, 39, 64, 82–83, 128, 176

analysis 100
behaviour 89
constraints 83
errors 85
of events 38
requirements 84
resource constraints 89
tolerable level of risk 44
tolerable risk 42
tool
   dynamic testing 191
   static testing 191
traceability 20, 66, 102, 112, 114, 121, 184
traceability matrix 113
transaction analysis 205

urn model 156
   single 164

validation 19, 25, 27, 31, 33, 37, 41, 102, 107–108, 125, 132, 136–137, 140–141, 176, 191, 207, 210
   empirical 166
   safety 52
verification 19, 25, 27, 31, 33, 37, 41, 84, 89, 102, 107, 125, 136–137, 140–141, 191, 207, 210–211
voting component 6
voting logic 62

walkthroughs 183
WCET 95
Weyukers axiom system 153
white box testing 31, 40, 104, 111–112, 147, 168
worst case execution times 3, 85, 94–95, 100

DATE DE RETOUR L.-Brault